Formulation and Stoichiometry

A REVIEW OF Fundamental Chemistry

Emil J. Margolis

THE CITY COLLEGE OF
THE CITY UNIVERSITY OF NEW YORK

Formulation and Stoichiometry

A REVIEW OF Fundamental Chemistry

APPLETON

CENTURY

CROFTS

New York

DIVISION OF
MEREDITH
CORPORATION

Copyright © 1968 by

MEREDITH CORPORATION

All rights reserved.
This book, or parts thereof, must not be used or reproduced in any manner without written permission. For information address the publisher, Appleton-Century-Crofts, Division of Meredith Corporation, 440 Park Avenue South, New York, N.Y. 10016

6107-1

Library of Congress Card Number: 67-28063

PRINTED IN THE UNITED STATES OF AMERICA

E 59795

PREFACE

The purpose of this book is to interpret more sensitively some of the offerings of the standard text book of general chemistry. As a supplement thereto, it covers various aspects of formulation and stoichiometry that are frequently treated far too perfunctorily or, in many instances, are not considered at all.

The inadequate attention often accorded by the comprehensive text to many topics within its proper purview arises, understandably enough, from the numerous broad and highly varied objectives set for the first year of the curriculum for modern chemistry in colleges and universities. For the serious student this means, more often than not, the frustrations of questions unanswered.

The amplification that this book proffers in the immediate area of its subject covers the equations representing *internal* redox reactions, not only of the simple but, also, of the multiple disproportionations of which the complexities often discourage an undertaking despite the challenge they offer: distinctions to be observed in the balancing of equations in contrasting alkali-basic and ammonia-basic reaction media; quantitative contributions made by the ionization or dissociation effects of electrolytes to the colligative properties of their solutions; intensive application of the universal reaction principle of *chemical equivalence* to the stoichiometry of oxidation and reduction.

As the endeavor here is always to anticipate the questions most likely to be asked by beginning students, it is deemed prudent not to start in the middle of things on the assumption, as often misconceived by supplemental texts, that what is needed in the way of preparation for comprehension has already been acquired elsewhere. Hence, various rudiments of theory normally taken for granted as understood are emphasized and elaborated whenever an efficacious integration of subsequent chemical themes must be ensured.

The fortifying of the text with significant, and frequently advanced, applications of formulation and stoichiometry establishes its utility not only to supplement a course, or for review, but also its character as a fully independent and self-sustaining primary text for instruction in class.

The varied exercises given—some "routine," others "challenging"— afford the student ample opportunities to demonstrate a confident understanding and practical mastery of principles, a memory for important

v

facts, and a capability with elementary mathematics. Indeed, they do more; for, by pedagogical design and physical construction, many of these exercises feature new chemical information not supplied in the text of the chapter. They thus offer earnest incentives to learn further while demonstrating skills in performance.

The ample numbers of worked-out problems, each interpreted in progressive step-by-step fashion, demonstrate practical methods of solving successfully the numerous mathematical exercises. Answers to all the exercises are fully supplied in the Appendix. Ready opportunity is thus given to corroborate having been on the right track.

E.J.M.

CONTENTS

CHAPTER

APPENDICES

CHAPTER 1

CONCEPTS OF QUANTITATIVE COMPOSITION

The objective of this book is to interpret the chemist's shorthand — chemical symbolism and formulation — and to introduce the subject of *stoichiometry*. Stoichiometry is the branch of chemical mathematics concerned with the relative amounts of substances that interact. Derived from the Greek *stoicheion,* meaning *element,* stoichiometry originally treated only of the different weights of elements that combine to form specific compounds. Modern usage, however, has expanded the generic applicability of the term to any and all aspects of chemical reactivity that can be validly interpreted by a balanced chemical equation. Consequently, stoichiometry encompasses within its purview not only weights but also volumes, gaseous and liquid, and even thermochemical and photochemical quanta of reaction.

The art and practice of calculating the combining weights of chemical elements requires a thorough knowledge of fundamental physical laws and theories.

LAW OF CONSERVATION — MATTER AND ENERGY

Mass-energy can be neither created nor destroyed. Although matter can be converted from one chemical species to another, energy transformed from one kind to another, and matter and energy mutually interconverted, no over-all losses or gains can occur in a closed system. This is the universal and incontrovertible law of conservation.

That mass and energy are equivalent and can be converted into each other was demonstrated in the classic theory of relativity developed by Albert Einstein in Germany in 1905. The interconvertibility of mass and energy is expressed mathematically as

$$E = mc^2,$$

where E represents the energy in ergs (fundamental unit of energy) obtained from the rest mass of a given quantity (in grams) of matter m that disappears; and c, the velocity of light (3×10^{10} centimeters a second).

1

Inherent in this mathematical definition is the recognition that the energy equivalent of any specific quantity of matter is dependent solely upon its total mass and not upon its chemical nature, form, or identity.

In *ordinary* types of chemical reaction — those that do not chemically involve the atomic nuclei — transformations of mass to energy are far too slight to be even detectable. Consequently, total mass remains unchanged in any nonnuclear reaction and the total initial mass of all reactants must equal the total final mass of all products.

MASS VERSUS WEIGHT

In practice the terms *mass* and *weight* are frequently confused. Despite their marked differentiation, they are loosely regarded as equivalent. Mass is, correctly, the *quantity* of matter ascertainable by the experimental measurement of "weighing" or by the response (change in velocity) that is caused by an unbalanced force.

In determining the mass of an object by weighing it upon a balance, we are actually measuring the magnitude of the force with which it is being attracted by the force of gravitation of the earth. As the force of gravity diminishes as distance from the earth increases, the *weight* of an object must likewise diminish. Mass, however, is completely independent of gravitational force; hence, although the weight of a given object decreases as the altitude of its measurement increases, its mass remains constant. An object is weighed by *counterbalancing* it with the exact number of *standard* units of mass (the "weights") required to make the gravitational force upon the object on one side equal to the gravitational force upon the standard weights on the other side. We are, in effect, canceling the force of gravitation, and thus we are measuring mass.

MASS AND VOLUME — THE IMPLICATIONS OF DENSITY

The term *volume*, as the chemist employs it, means precise, minimum amount of space needed to accommodate a specific amount of matter. The fundamental unit of volume in the metric system is the *cubic centimeter* (cc), which term is acceptably interchanged with *milliliter* (ml).

This duality of terms for volume emphasizes not only an error in computation that was originally made in the measurement of volume but also a difference in the units utilized in the respective measurements. Derived from units of *length*, the cubic centimeter represents the volume of a cube one centimeter on edge. An assignment of 1000 cc was made to standardize the space occupied by an intended mass of one kilogram of water at its temperature of maximum density, 3.98°C. It was subsequently determined, however, that one kilogram of the water actually occupies 1000.027 cc. To rectify this error without compounding a *fait accompli* into utter confusion, the term "liter" was designated to represent this volume of 1000.027 cc. Consequently, 0.001 liter (= 1 milliliter *exactly*) becomes identical in volume with 1.000027 cc. Hence, one gram of water at 3.98°C occupies a volume that may alternately be described as 1 ml or as 1.000027 cc. The quantitative difference between the two is so slight that there is little reason to belabor the inconsistency further; hence, the cc and the ml may be considered identical for all practical purposes.

Density is defined as the mass of a unit of volume. Using the metric units already established, the dimensions of density are grams/cc to the cubic centimeter (g/cc), or to the milliliter (g/ml). Density may also be expressed in any other units that properly relate the mass of a substance to the volume it occupies. This may be useful when experiments in weighing require the use of larger volumes, for example, in weighing gases, because reliably weighable quantities of materials so light must be, understandably, large. Thus, the densities of gases are often expressed in dimensions of gram/liter.

DENSITY VERSUS SPECIFIC GRAVITY

Closely related to density is *specific gravity* — a term of especial convenience when applied to liquids and solids. The specific gravity of a substance is a ratio of the density of that substance to the density of another substance taken as a standard. It thus represents how many times heavier the substance is than the stated standard. If, for instance, the standard is water at 3.98°C (more conveniently, 4°C) where its density is 1.0 g/ml, the specific gravity of the substance in question will be *numerically* equivalent to its density (*but* dimensionless, inasmuch as the units in the ratio will have cancelled out). Thus, for liquid mercury, of which the density is 13.6 g/ml,

$$\text{sp.gr.} = \frac{\text{density of mercury}}{\text{density of water at } 3.98°C} = \frac{13.6 \text{ g/ml}}{1.0 \text{ g/ml}} = 13.6.$$

Although a standard of 1.0 g/ml is a distinct convenience, laboratory practice rarely permits exactitude with respect to this, inasmuch as the temperature is seldom 3.98°C and liquids other than water may occasionally be preferred as standards. In any event, appropriate notation will clarify what is intended. The notation "$1.78^{25°}_{4°}$" specifies that the density of the given substance at 25°C is 1.78 times greater than that of water at 4°C and, consequently, the actual density of the given substance (at this temperature) must coincide with its specific gravity. Hence, density here is 1.78 g/ml.

Were the reference standard to have been water at 25°C, the specific gravity of the substance in question (expressed by sp. gr.$^{25°}_{25°}$) could not be numerically identical to its density, because the density of water at 25°C is something less than 1.0. The value of specific gravity for the given substance is now somewhat larger than the value of 1.78 computed for it when water at 4°C was used as standard — the value of the density ratio having increased because of a decrease in the denominator (the numerator remaining constant).

LAW OF DEFINITE PROPORTIONS—CONSTANCY OF COMPOSITION

Two or more substances — elements or compounds — which chemically combine to form a specific compound will do so in fixed and unalterable proportions by weight.

This scientific law, first expressed by Joseph Proust (France, 1799), means that the composition of each and every compound is incorruptibly exact and may always be defined in terms of the combining parts by weight of each of the elements therein. Certain contingencies that must, however, be recognized here are as follows:

1. Two elements may combine in different proportions by weight if each new ratio of combination leads to a completely different compound.

Thus, although hydrogen and oxygen, in forming the compound *water*, always combine in the ratio of *one* part by weight of hydrogen to approximately *eight* parts of oxygen — the same two elements may also combine in the ratio of *one* part by weight of hydrogen to *sixteen* parts of oxygen. But, the compound formed here is not water (H_2O); it is hydrogen peroxide (H_2O_2).

2. The Law of Definite Proportions does not preclude the possibility that the same elements may combine in identical proportions to form two or more different compounds. Such compounds are called *isomers*. Some isomers may be widely different in physical and chemical properties; for example, the compounds ammonium thiocyanate, empirically described by the molecular arrangement NH_4OCN, and urea, empirically described by the molecular arrangement $CO(NH_2)_2$.

The properties of other isomers may be so close that only a significant physical property offers an effective and ready means for differentiating them. Thus, *glucose* (grape sugar) and *fructose* (fruit sugar), both identified by the same general formula $C_6H_{12}O_6$, have virtually similar chemical and physical properties but they can be readily distinguished on the basis of their opposing optical activities. When polarized light (vibrations confined to a single plane) is transmitted through a solution of glucose, the plane of the polarized light is rotated to the right; when similar light is passed through a solution of fructose, however, rotation is to the left. The chemical synonyms "dextrose" and "levulose" are thus comprehendingly descriptive of the respective compounds.

3. Another type of chemically different compounds with exactly similar percentage composition are those constituted of precisely the same elements in numbers of atoms that are, in the different compounds, reducible to a common empirical (simplest) unit.

Thus, the compounds *formaldehyde*, CH_2O, and *glucose*, $C_6H_{12}O_6$, are both algebraically expressed by different numbers of units of CH_2O. In formaldehyde, only one such unit is present; for the glucose, six such units, formulated for the compound as $(CH_2O)_6$. Clearly, percentage composition, which expresses parts by weight of a constituent element per one hundred parts by weight of the entire compound, is not mathematically altered by the multiple that has been applied to the formula. The ratios C/CH_2O, $2H/CH_2O$, and O/CH_2O correspond exactly to $6(C)/6(CH_2O)$, $6(2H)/6(CH_2O)$, and $6O/6(CH_2O)$, respectively.

None of these variations compromises in the slightest the precise validity of confirmed law, because percentage composition by weight remains incorruptibly constant for each and every specific compound. When proportions of the elements are the same for different compounds, the numbers and architectural arrangements of their atoms will supply the answers to the different properties.

Another seeming refutation of the Law of Definite Proportions must be reconciled here. References heretofore have been made to naturally occurring substances. In such samples of matter it must be recognized that the percentage distributions of the various *isotopes* (different varieties of atoms of the same element) remain uniform under all normal conditions. It is clear that if one particular isotope of a given element were combined with some single specific isotope of another the analysis might well show a percentage composition for the compound formed that is significantly different from the one shown under reaction conditions of normal mixtures of the isotopes of the element.

In illustration, we have observed that one part of hydrogen will combine with eight parts of oxygen to form water — an analysis reflecting the distributive percentage *average* atomic weight of the three isotopic atoms of natural hydrogen (mass numbers 1, 2, and 3), and of the three isotopic atoms of natural oxygen (mass numbers 16, 17, and 18). Were we to have taken, however, specifically individual isotopes of the respective elements we would have obtained "H_2O" of different compositions by weight, in conformity with the possibilities shown in Table 1:1.

Molecules of "heavy water," D_2O, and the still heavier T_2O, all contribute to the average weight of molecules of water. The latter are standardized in the numerical identity of the ratios of elemental weights solely

TABLE 1:1

Possible Combinations of Hydrogen and Oxygen

Isotopic Combinations	H_2O: *hydrogen to oxygen weight ratios*
Protium ($_1H^1$) oxides $\begin{cases} _1H^1 \text{ with } _8O^{16} \\ _1H^1 \text{ with } _8O^{17} \\ _1H^1 \text{ with } _8O^{18} \end{cases}$	$2:16$ or $1:8$ $2:17$ or $1:8\frac{1}{2}$ $2:18$ or $1:9$
Deuterium ($_1H^2$ or D) oxides $\begin{cases} _1H^2 \text{ with } _8O^{16} \\ _1H^2 \text{ with } _8O^{17} \\ _1H^2 \text{ with } _8O^{18} \end{cases}$	$4:16$ or $1:4$ $4:17$ or $1:4\frac{1}{4}$ $4:18$ or $1:4\frac{1}{2}$
Tritium ($_1H^3$ or T) oxides $\begin{cases} _1H^3 \text{ with } _8O^{16} \\ _1H^3 \text{ with } _1O^{17} \\ _1H^3 \text{ with } _8O^{18} \end{cases}$	$6:16$ or $1:2\frac{2}{3}$ $6:17$ or $1:2\frac{5}{6}$ $6:18$ or $1:3$

because all samples contain virtually constant percentages of admixed H_2O, D_2O, and T_2O. Although the reference to *heavy water* is, in general, intended to distinguish between hydrogen atoms — the protium and the deuterium or tritium — in their chemical associations with the oxygen atom of average weight, the same considerations apply with respect to the isotopic variations that are possible with the oxygen atoms in their chemical associations with a hydrogenation of *average* weight. The abundance of the protium, to the near-exclusion of the other varieties, ensures a practical average atomic weight of hydrogen of 1.

LAW OF MULTIPLE PROPORTIONS — PRELUDE TO ATOMIC CONCEPTS

Whenever variable weights of one specific element combine with a fixed weight of another specific element in the formation of different compounds of the two, those variable weights will always be found to be simple multiples of one another.

By a "simple multiple" is meant a number expressible by a small whole-number numerator and a small whole-number denominator; for example, $\frac{1}{2}$, $\frac{3}{1}$ ($= 3$), $\frac{2}{3}$, etc.

The interpretation of this scientific law (first stated by John Dalton, England, in 1803) leads inevitably to the conclusion that in ordinary chemical reaction, whole atoms are nature's building blocks which must be transferred intact from one substance to another. Any weight of any substance must then represent an integral number of whole atoms, and any removals or additions in the formation of new substances must likewise be in the ratios of integral numbers of whole atoms. These integral numbers then establish the relative numbers of each of the pertinent atoms present in each of the respective compounds.

ATOMIC THEORY — IMPLICATIONS OF MULTIPLE PROPORTIONS

John Dalton's interpretations of multiple combining proportions led him to propose a theory of internal constitution of matter that has served as the foundation of our modern developments of atomic concept. Its postulations were, essentially, as follows: (we interpolate certain modernized appraisals):

1. The chemical elements and all the compounds formed from their combinations are composed of discrete and indivisible entities called *atoms*. [The existence of *subatomic* particles — not surmised in Dalton's time — must today represent a qualification of the term "indivisible." For all ordinary (nonnuclear) chemical changes, however, the term remains legitimate.]

2. The chemical and physical properties of all atoms of any specific element are identical, but are different from those of any other element. (Here again the existence of isotopes must qualify the term "identical," at least insofar as identity of the physical property of mass is concerned.)

TABLE 1:2

Different Compounds of Nitrogen and Oxygen

Compound	Parts by Weight		Corresponding Ratios	Molecular Formula
	Nitrogen (constant)	Oxygen and the variable Simple Multiples	N : O in Whole Atomic Weights	
Nitrous Oxide	14.00	$8.00 \left(= \frac{1}{2} \times 16.00\right)$	$(2 \times 14.00):(1 \times 16.00)$	N_2O
Nitric Oxide	14.00	$16.00 \left(= \frac{1}{1} \times 16.00\right)$	$(1 \times 14.00):(1 \times 16.00)$	NO
Nitrogen Trioxide	14.00	$24.00 \left(= \frac{3}{2} \times 16.00\right)$	$(2 \times 14.00):(3 \times 16.00)$	N_2O_3
Nitrogen Dioxide	14.00	$32.00 \left(= \frac{2}{1} \times 16.00\right)$	$(1 \times 14.00):(2 \times 16.00)$	NO_2
Nitrogen Pentoxide	14.00	$40.00 \left(= \frac{5}{2} \times 16.00\right)$	$(2 \times 14.00):(5 \times 16.00)$	N_2O_5

3. Chemical change involves merely the combinations of different elements to form compounds; and the atoms of the elements involved in such change are merely rearranged without impairment of their identities. (We must qualify "chemical change" as used here as ordinary, or non-nuclear.)

4. The chemical combinations of different atoms that lead to the formation of compounds must occur in simple numerical ratios of whole numbers; that is, 1 : 1, 2 : 1, 2 : 3, etc.).

Let us now illustrate the implications of multiple proportions. In Tables 1:2 and 1:3, the parts-by-weight base for each weight-constant element is expressed in grams; and these, moreover, have been conveniently rounded off for greater visual clarity. It should be borne in mind that the weights represent a judicious refining of experimental, and hence, sensitivity-limited, measurements.

The identical results could have been achieved with equal convenience by making oxygen the element of constant weight, and observing the variations in weight of the nitrogen; and similarly relating the simple multiples to units of whole atomic weights, as shown in Table 1:3.

TABLE 1:3

Alternate Calculations for Oxides of Nitrogen

	Parts by Weight		Corresponding Ratios	
Compound	Nitrogen and the variable Simple Multiples	Oxygen (constant)	N : O in Whole Atomic Weights	Molecular Formula
Nitrous Oxide	$28.00 \left(= \frac{2}{1} \times 14.00 \right)$	16.00	$(2 \times 14.00) : (1 \times 16.00)$	N_2O
Nitric Oxide	$14.00 \left(= \frac{1}{1} \times 14.00 \right)$	16.00	$(1 \times 14.00) : (1 \times 16.00)$	NO
Nitrogen Trioxide	$9.33 \left(= \frac{2}{3} \times 14.00 \right)$	16.00	$(2 \times 14.00) : (3 \times 16.00)$	N_2O_3
Nitrogen Dioxide	$7.00 \left(= \frac{1}{2} \times 14.00 \right)$	16.00	$(1 \times 14.00) : (2 \times 16.00)$	NO_2
Nitrogen Pentoxide	$5.60 \left(= \frac{2}{5} \times 14.00 \right)$	16.00	$(2 \times 14.00) : (5 \times 16.00)$	N_2O_5

As would be mathematically inevitable, the two simple multiples obtained for any specific individual compound, when the weight of one element therein is alternately varied against a fixed weight of the other, are *reciprocally* related to one another.

GAY-LUSSAC'S LAW — WHOLE-NUMBER RELATIONSHIPS

Dalton's atomic theory suggested no clues whatever as to the manner in which the relative weights of atoms might be ascertained. The implications of the integers in the Law of Multiple Proportions not only were to prove inescapable but also, indeed, were to receive additional bolstering from an entirely different experimental direction — again, by the recurrence of integers.

In 1808, J. L. Gay-Lussac (France) observed that the volumes of all gases that combine or are produced in a chemical reaction may always be expressed in ratios of small whole numbers.

This statement, in various forms of its intent, is today called the Law of Combining Volumes. Although their theoretical significance was not understood at the time, the known experimental phenomena proved later to be additional clues to what was being sought; namely, a scale of atomic weight.

A few examples demonstrate the Law of Combining Volumes:

(a) $\quad$ nitrogen $\quad + \quad$ hydrogen $\quad \rightarrow \quad$ ammonia
$\quad$ 1 volume, gas $\quad$ 3 volumes, gas $\quad$ 2 volumes, gas

$$\text{ratio} = 1 : 3 : 2$$

(b) $\quad$ ethane $\quad + \quad$ oxygen $\quad \rightarrow \quad$ carbon dioxide $\quad + \quad$ water vapor
$\quad$ 2 volumes, gas $\quad$ 7 volumes, gas $\quad$ 4 volumes, gas $\quad$ 6 volumes, gas

$$\text{ratio} = 2 : 7 : 4 : 6$$

(c) $\quad$ hydrogen $\quad + \quad$ chlorine $\quad \rightarrow \quad$ hydrogen chloride
$\quad$ 1 volume, gas $\quad$ 1 volume, gas $\quad$ 2 volumes, gas

$$\text{ratio} = 1 : 1 : 2$$

(d) $\quad$ carbon $\quad + \quad$ oxygen $\quad \rightarrow \quad$ carbon dioxide
$\quad$ solid $\quad$ 1 volume, gas $\quad$ 1 volume, gas

$$\text{ratio} = 1 : 1$$

It is to be observed that, unlike the "left hand/right hand" equalities that represent conservation of mass, there is no conservation of volume.

The relationships witnessed hold regardless of the particular units used to express volume — liters, cubic centimeters, cubic feet, etc. — provided that they are uniformly applied to a given reaction. Likewise, it must be evident that any expression of a relationship in fractional fashion — such as $(\frac{1}{3}) : 1 : (\frac{2}{3})$, as might be written for example a — does not preclude its being translated into integral numbers.

AVOGADRO'S PRINCIPLE

In 1811, Amadeo Avogadro (Italy) offered an explanation of the yet-to-be-understood theoretical significance of combining volumes of gases. He hypothesized that

> equal volumes of all gases under the same conditions of temperature and pressure contain equal numbers of molecules.

Through lack of experimental evidence to support this contention it was not accepted at the time, and as a result the concept lay dormant for about fifty years — when experimental findings in newly developed areas of physical and organic chemistry exhaustively corroborated it as valid. It has earned, today, the dignity not merely of an educated guess — an hypothesis — but rather, of a firmly established scientific law.

Let us examine, by means of a few examples, the theoretical implications of this concept and the historical objections that it had to surmount. It must be remembered that Dalton conceived of an elementary gas as a collection of solitary monoatomic entities, rather than as the polyatomic units (in virtually all ordinary instances) that today we call *molecules*. Dalton's inability to accept the theoretical premises of Gay-Lussac's laboratory investigations and of Avogadro's interpretations rested upon grounds that then seemed plausible. If in the reaction

$$\text{hydrogen gas} + \text{chlorine gas} \rightarrow \text{hydrogen chloride gas}$$

the experimental facts are

$$1 \text{ volume} + 1 \text{ volume} \rightarrow 2 \text{ volumes}$$

then, in accordance with Dalton's stipulations that each gas must be composed of indivisible atoms, it would follow that

$$1 \text{ atom} + 1 \text{ atom} \rightarrow 2 \text{ molecules}.$$

This last is obviously an impossibility if we are to accept the atom as the fundamental indivisible and independent unit of an elementary gas, because one indivisible unit combining with another indivisible unit could not yield more than one molecule. The carry-over of Dalton's single-atom concept of gases has left us, even up to the present time, with an arbitrary standard of weight combinations of an element — called its *combining weight* (or *equivalent weight*). This stemmed from the theoretically erroneous conclusion that because 1 part by weight of hydrogen combines with 8 parts by weight of oxygen to form water vapor, consequently, with the atomic weight of hydrogen fixed at 1, the relative atomic weight of oxygen should then be 8, and the Dalton equation for the reaction would be $H + O \rightarrow HO$.

As oxygen is such a prolific combiner with other elements, it seemed expedient then to define the proportions by weight of elements combining to form compounds in terms of the quantities that react with 8.0000 parts by weight of oxygen. This proportion by weight of hydrogen to oxygen is by no means changed by our present knowledge that both hydrogen gas and oxygen gas consist not of monoatomic units but of diatomic units — which can be divided into single atoms. Though the atomic weight of oxygen is consequently 16, the hydrogen-oxygen relationship of 2 : 16 in the formation of water vapor is still the proportion of 1 : 8. Complete consistency with the original intent that the combining weight of the element be that weight that combines with one atomic weight of oxygen would dictate the assigning to it of 16.0000, rather than 8.0000, parts by weight of oxygen.

But, hindsight is always 20-20 vision, to be sure. As a matter of fact, the revisions, in 1961, of the scale of atomic weights, wherein the basis for atomic weights was transferred from oxygen to carbon, would necessitate, from the point of view of complete consistency with original intent, a standard for combining weight not of 8.0000 parts by weight of oxygen, but rather, of 3.0000 parts by weight of the carbon-12 isotope.

Returning to the aforementioned reaction,

1 volume hydrogen + 1 volume chlorine → 2 volumes hydrogen chloride.

Any theory that did not require the splitting of an atom had to be founded on the assumption that the particulate units of each of the reacting gases had to be polyatomic entities; and this was the hypothesis advanced by Avogadro. Hence, the equation for the reaction in terms of such a concept would be not the impossible

1 atom H + 1 atom Cl → 2 molecules HCl,

but rather the quite credible

1 molecule H_2 + 1 molecule Cl_2 → 2 molecules HCl.

With equal validity, inasmuch as all that has been established by this procedure is the need for an *even* number of atoms in each of the hydrogen and chlorine molecules that interact, we might have written

1 molecule H_{2n} + 1 molecule Cl_{2n} → 2 molecules H_nCl_n,

with n representing some small whole number (2, 3, or 4, etc.).

CANNIZZARO'S PRINCIPLE

For nearly 50 years after Avogadro postulated (1811) that equal volumes of all gases under similar conditions of temperature and pressure contain the same numbers of molecules, much perplexity prevailed as to what was to be regarded as a workable scale of relative atomic weights. These difficulties reflected, in themselves, a lack of clear distinction between a molecule (and its gram-molecular weight) and an atom (and its gram-atomic weight). Of the conflicts between the Dalton concept of the gaseous molecule and the experimental relationships of combining volumes, we have already made mention. In the absence of experimental data, the suggestions of the polyatomic character of gaseous elements (inherent in the Avogadro hypothesis) proved something less than persuasive.

In 1858, however, the dormant Avogadro hypothesis was revived by Stanislao Cannizzaro (a former pupil of Avogadro) to interpret validly certain experimental observations that had just been made. At long last there appeared a solution of the dilemma of atomic and molecular weights and of the formulas to which they conformed. Cannizzaro's logical extending of the Avogadro hypothesis to a systematic determination of approximate atomic weights actually proved to be the means to confident acceptance by chemists of an atomic postulations theory.

Cannizzaro's conclusions were essentially as follows:

1. Every molecule is composed of an integral number of atoms. Therefore, every molecular weight must be comprised of an integral number of atomic weights. One gram-molecular weight of a compound must contain, consequently, either one gram-atomic weight of each of its constituent elements or some whole-number multiple thereof.

2. Inasmuch as *one* is the very smallest number of atoms that any molecule of a compound can contain, it is to be anticipated that an examination of a sufficient number of compounds of that element must eventually reveal one that contains just a single atom of that element. This, then provides the gram-atomic weight of that element.

3. To obtain the weight of the single atom, the relative weight of the single molecule is also needed. If, under similar conditions, identical volumes of gases contain equal numbers of molecules, then the ratio of the weights of these equal volumes of gases must be identical with the ratio of the weights of the comprising individual molecules. This provides a scale of relative molecular weights of various gaseous compounds.

4. A "standard" of molecular weight must be arbitrarily defined, against which the values of weights of other molecules, elemental or compound, may be fixed. If a value of 1 be assigned to the atom in the molecule of the very lightest gas known — hydrogen — then determining the number of atoms in this gaseous element will permit it to be used as a standard of molecular weight; that is, its molecular weight must be some integral multiple of 1. The necessary elucidations of this molecular weight are provided by the experimental measurements of the volumes of gases that chemically interact. Among others is the following reaction in terms of what has already been presented:

Reaction: hydrogen gas $+$ chlorine gas $\rightarrow$ hydrogen chloride gas

Volume measurement: 1 volume $+$ 1 volume $\rightarrow$ 2 volumes

Interpretation: 1 molecule $+$ 1 molecule $\rightarrow$ 2 molecules

n molecules $+$ n molecules $\rightarrow$ $2n$ molecules.

As 2 volumes, and its proportionate 2 molecules, of hydrogen chloride are being obtained, clearly there must be an even number of atoms in a molecule of hydrogen gas — for, we cannot settle for less than one atom of hydrogen in each of the two molecules of hydrogen chloride produced. These are merely relative ratios, however; and so, we really do not know from this whether the even number is exactly 2, or 4, or 6, etc. As Cannizzaro chose to fix the atomic weight of the hydrogen atom as 1, the empirical weight of the hydrogen unit in its gaseous reaction form is clearly 2.

5. If the empirical weight of the gaseous hydrogen unit, 2, be now regarded as the weight of the actual molecule in its true and full composition — its molecular weight — it becomes possible to calculate the atomic weights of other gaseous elements which, in reaction, yield gaseous products. In each instance we start with knowing the density of the gas, relative to hydrogen. If we were to determine the approximate atomic weight of oxygen, for example, we would measure the densities of a number of gaseous oxygen-containing compounds; or of the pure gaseous element itself, relative to that of hydrogen gas under similar conditions. Thus, if pure elemental oxygen is found to be 16 times more dense than hydrogen gas on the scale, $H_2 = 2$, the weight of each molecule of oxygen is 32. Further, were the density of water vapor found to be 9 times greater than that of hydrogen gas under the same conditions, its molecular weight would be 18.

The reaction between hydrogen and oxygen gases at high temperature to form water vapor may be interpreted with respect to the clues that it offers about the distributions of oxygen atoms and of the possibilities of ascertaining therefrom the atomic weight of oxygen. Thus,

Reaction: hydrogen gas $+$ oxygen gas $\rightarrow$ water vapor $(= H_2O)$

Volume measurements: 2 volumes $+$ 1 volume $\rightarrow$ 2 volumes

Interpretations: 2 molecules $+$ 1 molecule $\rightarrow$ 2 molecules

$2n$ molecules $+$ n molecules $\rightarrow$ $2n$ molecules.

Manifestly, if the hydrogen molecule is diatomic, both atoms of each of the two H_2 molecules must find their way into, and appear exclusively within, a separate molecule of water vapor. We have recorded the density of oxygen gas as 16 times greater than that of hydrogen gas; consequently, its molecular weight relative to that of hydrogen is 32. Inasmuch as only 1 molecule of oxygen gas is required for the formation of 2 molecules of water vapor, it is clear that just one-half of this weight, 16 units, must appear in each of the two separate molecules of water vapor. On this basis alone we cannot conclude that the atomic weight of oxygen is 16, for all that has been established here is that one-half of all of the oxygen atoms in oxygen gas have been chemically transferred to a single H_2O molecule. Were it known that the oxygen molecule is diatomic, we could conclude validly that the atomic weight of oxygen is 16. But, if all that is known is that half of some even number of oxygen atoms has been chemically transferred, then if oxygen molecules were tetra-, hexa-, or octa-atomic, etc., the "one-half of 32" which we have taken as the transferred weight would represent not 1 oxygen atom but, respectively, of 2, 3, or 4, etc.

Modern instrumental methods have indisputably established that the oxygen molecule is diatomic, O_2, with consequent assignment of 16 mass units to the average, naturally occurring, oxygen atom. Cannizzaro's analysis was necessarily dependent upon the limited equipment available in his day. His experimental procedure for evaluating the atomic weight of an element was as follows:

1. Determinations of the densities of gases containing the pertinent element, on the scale "$H_2 = 2$," under identical conditions of temperature and pressure. This gave the molecular weight of each of the gases.

2. Chemical analysis for the weight of the pertinent element contained in one gram-molecular weight of each of the various gases in which it appeared. This fractional weight of the whole gram-molecular weight of the compound represented the total weight of all of the atoms of the respective element that were present.

3. Correlation of the derived information permitted the extraction of the smallest weight of the respective element in analysis of sufficient numbers of different gaseous compounds in which it appeared. This smallest weight represented the atomic weight of the respective element.

The basis of justification for the approximate atomic weights — $O = 16$, $C = 12$, $Cl = 35.5$, $N = 14$ — may be observed even from the quite sparse data shown in Table 1:4.

It is to be conceded that the Cannizzaro method did not protect against the possibility that this "smallest weight" might prove at some later date to be an integral multiple of something still smaller in a newly discovered compound. The likelihood, however, of finding, for example, an 8 for oxygen, or a 7 for nitrogen, or a 6 for carbon, would appear to be quite remote, indeed. For, not only have analyses of percentage composition failed to turn up such elemental masses for the respective atoms but also, none of the numerous compounds already known show the odd-number multiples that might justify these smaller values.

Oxygen, for example, has yielded neither a "24" (3×8) nor a "40" (5×8). Nitrogen compounds have revealed neither a "21" (3×7) nor a

TABLE 1:4

Basic Calculations for Approximate Atomic Weights

Gaseous Compound	Gram-Molecular Weight	Grams of Element Per Gram-Molecular Weight				
		Hydrogen	Oxygen	Carbon	Chlorine	Nitrogen
Benzene	78.0	6.0	—	72.0	—	—
Carbon Monoxide	28.0	—	16.0	12.0	—	—
Carbon Dioxide	44.0	—	32.0	12.0	—	—
Ethane	30.0	6.0	—	24.0	—	—
Methane	16.0	4.0	—	12.0	—	—
Butane	58.0	4.0	—	48.0	—	—
Formic Acid	46.0	2.0	32.0	12.0	—	—
Ammonia	17.0	3.0	—	—	—	14.0
Nitrous Oxide	44.0	—	16.0	—	—	28.0
Nitric Oxide	30.0	—	16.0	—	—	14.0
Nitrogen Dioxide	46.0	—	32.0	—	—	14.0
Water (vapor)	18.0	2.0	16.0	—	—	—
Chloroform	119.5	1.0	—	12.0	106.5	—
Hydrogen Chloride	36.5	1.0	—	—	35.5	—
Carbon Tetrachloride	154.0	—	—	12.0	142.0	—
Methyl Chloride	50.5	3.0	—	12.0	35.5	—

"35" (5 × 7). And there is no "18" (3 × 6) or "30" (5 × 6) for carbon in any of the many hundreds of thousands of its known compounds. The only way that such reduced values might plausibly be supposed to "exist" despite the present lack of any experimental evidence for their support would be to assume that all compounds of oxygen, nitrogen, or carbon are necessarily composed exclusively of even numbers of atoms. However, such a premise would appear to be completely demolished by present-day concepts of atomic and molecular structure.

LAW OF DULONG AND PETIT—HEAT CAPACITY AND ATOMIC WEIGHTS

As the Cannizzaro method of approximating atomic weights is restricted to elements that can form several volatile gaseous compounds, it poses the question of what to do about elements that do not. Solid elements, in particular, form only limited numbers of volatile compounds, if any, so the Cannizzaro procedure cannot be applied to them. In 1819, however, Pierre L. Dulong and Alexis Petit had postulated a principle concerning the capacities of solid elements to absorb heat which now proves useful in making reasonably close estimates of atomic weights of the solid elements. Coupling these estimates with other available experimental data permits an accurate placement of the exact atomic weight.

Dulong and Petit observed that the heat capacity of a gram-atom of a metal in its elemental state was *constant* at a value of approximately 6 calories for each degree centigrade of change in temperature. This value is frequently called the *atomic heat constant*. As each gram-atom of any element contains 6.023×10^{23} atoms — the Avogadro number — the significance of the observation of Dulong and Petit is that the capacity of an atom to absorb heat is independent of its specific chemical identity; and that equal quantities of heat will be required to raise identical numbers of atoms through similar intervals of change in temperature. As we work experimentally with weights in grams, and inasmuch as one gram of one element does not contain the same number of atoms as one gram of another (distinction of "gram" from "gram-atom" or "mole"), the heat required to raise equal numbers of grams of different elements through equal intervals of temperature will be different.

We refer to heat capacity of a gram as *specific heat* — the number of calories, that is, required to raise the temperature of one gram of the solid element by one degree centigrade. In terms, then, of gram-atomic weight and specific heat we may equate as follows:

$$\frac{\text{atomic weight}}{\text{(grams/gram-atom)}} \times \frac{\text{specific heat}}{\text{(cal/}^\circ\text{C gram)}} = \frac{6}{\text{(cal/}^\circ\text{C gram-atom)}}$$

The empirical approximation that must be made of the atomic heat constant will be apparent from the typical figures given in Table 1:5. These show not only that the evaluations for lighter elements differ widely from those made for heavier elements at ordinary temperatures, but also that at very low temperatures the heat capacities of all substances drop sharply. The interpretation of Dulong and Petit, that the multiplication product of atomic weight and specific heat is a constant, actually belies the experimental evidence that molar heat capacities of all elements increase as temperature increases. Hence, in reality, atomic heat constants are specific evaluations for specific temperatures. At approximate room temperatures, however, we are on fairly common ground inasmuch as a value of 6 represents the limiting value of atomic heat capacity approached by most elements.

By a judicious exclusion of some elements — particularly the very lightest, such as beryllium, boron, and carbon — the heat capacities of the heavier metals at ordinary temperatures may be refined to an average of 6.2-6.4 calories per degree per mole.

Let us now pursue a specific application of the procedure of Dulong and Petit wherein the derived localization of the neighborhood of correct atomic weight permits a chemical analysis of composition by weight which can be used in conjunction with it to determine the precise atomic weight. Laboratory analysis of a sample of copper oxide reveals that 79.89% of the total weight of any sample of the compound is copper, and that the balance,

TABLE 1:5

Elemental Heat Capacities and Atomic Weights

Element	Atomic Weight (grams/gram-atom)	Specific Heat (cal/gram-degree)	Atomic Heat Constant (cal/mole-degree)
Carbon (diamond)	12.0	0.12	1.4
Carbon (graphite)	12.0	0.17	2.0
Boron	10.8	0.31	3.3
Beryllium	9.01	0.39	3.5
Aluminum	27.0	0.21	5.7
Magnesium	24.3	0.24	5.8
Copper	63.5	0.092	5.8
Sulfur	32.1	0.18	5.8
Antimony	122	0.050	6.1
Arsenic	74.9	0.082	6.1
Iron	55.8	0.11	6.1
Gold	197	0.031	6.1
Zinc	65.4	0.093	6.1
Tin	119	0.054	6.4
Lead	207	0.031	6.4
Silver	108	0.056	6.4
Nickel	58.7	0.11	6.5
Lithium	6.94	0.94	6.5
Iodine	127	0.052	6.6
Uranium	238	0.028	6.7
Potassium	39.1	0.18	7.0

20.11%, is the weight of the oxygen. Consequently, in 1.000 g of copper oxide compound,

$$\text{copper} = 0.79789 \times 1.000 = 0.7989 \text{ g Cu}$$
$$\text{oxygen} = 0.2111 \ \times 1.000 = 0.2011 \text{ g O.}$$

In terms of the weight of copper that has combined with 15.9994 g of oxygen (its gram-atomic weight), we derive

0.211 g O combines with 0.7989 g Cu

1 g O combines with 0.7989/0.2011 g Cu

15.9994 g O combine with (0.7989/0.2011) $\times$ 15.9994 g Cu.

which yields

1 gram-atomic weight of O combines with 63.54 g Cu.

Now, does this weight of 63.54 g represent the gram-atomic weight of Cu; or is it an integral multiple, or even a submultiple, of it? Were each molecule of this compound composed solely of one atom of copper, clearly the atomic weight of this element must be 63.54 (*atomic mass unit*) (amu), corresponding to 63.54 g — the gram-atomic weight or mole weight, that of 6.023×10^{23} copper atoms.

Were the molecule composed of 2, or 3, or n atoms on the other hand, just as manifest would be our need to divide 63.54 amu by 2, or 3, or n in order to obtain the mass of a single atom. The number of Cu atoms in a molecule remains unrevealed by the present information; yet, the Dulong and Petit principle, despite its approximate nature, can determine whether it is a multiple or submultiple of the true atomic weight. In either instance, the integer would be clearly discernible, mathematically. Using the specific heat given in Table 1:5 and employing the commonly accepted average of 6.2 for the atomic heat constant (anything from 6.1 to 6.4 would do just as well), we derive the following for the Cu:

$$\text{atomic weight} = \frac{6.2 \text{ cal/gram-atom degree}}{0.092 \text{ cal/gram degree}}$$

$$= 67 \text{ g/atom (approx)}.$$

It is obvious from this that 63.54 g must be the gram-atomic weight of Cu, for 63.54 is as close as we can reasonably come to the Dulong and Petit estimate of 67 by multiplying or dividing by a *whole* number; for there must be *whole* numbers of atoms. That is, with $63.54 \times 1 \approx 67$, the true atomic weight of Cu is, consequently, 63.54 amu, and we therefore assign a formula of CuO to the compound in which it appeared.

Another brief example will serve usefully to confirm the Dulong and Petit rule. Just before it was postulated, the proposal had been made — strictly on the grounds of percentage-weight analysis—that the gram-atom of silver be assigned a weight of 432.5 g. Let us examine the validity of such an assignment in the light of subsequent developments. The table reveals a value of 0.056 cal/gram degree for the specific heat of silver. Therefore, for silver

$$\text{atomic weight} = \frac{6.2 \text{ cal/gram-atom degree}}{0.056 \text{ cal/gram-degree}}$$

$$= 110 \text{ g/atom (approx)}.$$

Now, we would have to divide the 432.5 by 4 in order to obtain a weight in the immediate vicinity of this close estimate (110) of the true atomic weight. The present-day establishment of the atomic weight of silver at 107.870 ± 0.003 amu ($= 107.870 \pm 0.003$ g/gram-atom) reflects not only the validity of the Dulong and Petit rule and its usefulness but also the refinements of experimental techniques and apparatus in acquiring more precise measurements of composition by weight.

It should be pointed out that as a number of solid elements can be volatilized into gaseous compounds, the opportunity was afforded to check the validity of the Dulong and Petit principle by the Cannizzaro method.

THE MOLE CONCEPT — THE AVOGADRO NUMBER (N)

The computations of the quantities of substances that chemically interact, and of the relative numbers of particles that such weights represent, are highly facilitated by employing the unit of quantity termed the *mole*.

The resemblance of this term to "molecule" inevitably conveys the original intent with which the concept of the weight of the mole was applied. Modern usage, however, now quite broadly applies the concept of the mole to any class of substance, elemental or compound, and regardless of whether it be in atomic, or molecular, or simple or complex polyatomic ionic condition.

The mole is the weight of any substance which, expressed in grams, is numerically identical with the number of atomic mass units (amu) that represents the weight of any single unit particle or individual entity of the substance. Illustratively, bearing in mind that the total mass of any particle — simple or polyatomic — is the sum of the weights of the atoms that comprise it, we determine that:

1 mole of H_2O molecules weighs 18.0143 g, but 1 molecule of H_2O has a mass of 18.0143 amu

1 mole of Cl_2 molecules weighs 70.906 g, but 1 molecule of Cl_2 has a mass of 70.906 amu

1 mole of Hg_2^{2+} ions weighs 401.18 g, but 1 ion of Hg_2^{2+} has a mass of 401.18 amu

By analogy, 1 mole of Cu atoms, weighing 63.54 g, may chemically be converted under different conditions into 63.54 g of Cu^+ ions, or Cu^{2+} ions, with the conversion of 1 atom of Cu of mass 63.54 amu. In the alternative conversions, it would yield a single Cu^+ ion of mass 63.54 amu, or a single Cu^{2+} ion of mass 63.54 amu.

Inherent in these relationships is the implication that equal numbers of moles of all substances contain virtually identical *numbers* of comprising independent unit particles, regardless of differences in chemical identity. Several different and unrelated experimental procedures have provided ample and confident corroboration of this principle. Indeed, the various techniques of investigation are sensitive enough to define with remarkable agreement the actual numbers of atoms, molecules, or ions present in the mole of the given substance — 6.023×10^{23}. This is numerically identified with the letter N and called the Avogadro number for the Italian physicist who hypothesized that the volumes of gases indicate the numbers rather than the weights of the molecules therein.

The fundamental importance of this physicochemical constant merits some exposition of the manner in which it was derived. Avogadro himself had no way of knowing it and the half-century following Avogadro's work yielded no reliable values of N. The complicated theoretical involvements of the highly diverse and independent experimental approaches by which remarkably concurrent values of N have been achieved naturally suggest judicious restraint in description here. It suffices to point out that the instrument known as the *mass spectrometer* allows the mass of a particle to be determined with a high degree of accuracy. The average mass of the helium molecule is thus closely evaluated at 6.66×10^{-24} gram. To simplify our computations with such extremely small masses we have adopted the *atomic mass unit* (amu) — with mass equated as

$$1 \text{ amu} = 1.661 \times 10^{-24} \text{ g.}$$

Consequently, the mass of the helium molecule, in amu,

$$= \frac{6.66 \times 10^{-24} \text{ g/molecule}}{1.660 \times 10^{-24} \text{ g/amu}}$$

$$= 4.00 \text{ amu/molecule.}$$

In accordance with our definition of "mole," the molecular weight of helium is, synonymously, 4.00 g/g-molecule. Now, how many molecules are actually present in each gram of the helium? By definition, as every gram of the total weight of the gram-molecule may be represented by a mass of 1 amu ($= 1.660 \times 10^{24}$ g) of the total mass of a single molecule, the 4.00 g/g-molecule, which represents the molecular weight of helium, must likewise represent

$$\frac{4.00 \text{ g/g-molecule}}{(4.00 \times 1.660 \times 10^{-24} \text{ g})/1 \text{ molecule}}$$

$$\approx 6.02 \times 10^{23} \text{ molecules/g-molecule.}$$

The significance of the number 6.02×10^{23} is apparent both from the observation that molecular weight cancels out from numerator and denominator, and from the applicability of the arithmetic to similar determinations of the numbers of atoms in a gram-atom, the numbers of ions in a gram-ion; or of even the numbers of electrons in a gram-electron. The value of N is always 6.02×10^{23} particles of the substance of the mole, provided that all mass relationships are stated in grams and atomic mass units.

In the light of what has preceded, refinements of terminology are in order, not only with respect to the dual connotation of the word *mole*, applied to both the number and the weight of particle, but also with respect to various alternative terms that are still in fairly common use and which already have been employed in certain of the expositions we have given. The following tabulations reveal the exact distinctions to be made (but, admittedly, not always followed) on the basis of the specific particles being identified by the "mole":

$$\text{mole} = \text{mole (or molar) weight} \begin{cases} \text{gram-atomic weight} \\ \text{gram-molecular weight} \\ \text{gram-ionic weight} \end{cases} \begin{array}{c} \text{gram-formula} \\ \text{weight} \end{array}$$

$$\text{mole} = 6.02 \times 10^{23} \text{ unit particles} \begin{cases} \text{gram-atom} \\ \text{gram-molecule} \\ \text{gram-ion} \\ \text{gram-electron, or faraday unit of} \\ \quad \text{electricity} \end{cases} \text{gram-formula}$$

number of moles

$$= \frac{\text{total weight in grams (of atom, molecule, or ion)}}{\text{gram-formula weight or mole weight (of atom, molecule, or ion)}}$$

Illustratively, the numbers of moles represented by (a) 50.0 g of silver atoms; (b) 50.0 g of copper ions; and (c) 50.0 g of carbon dioxide molecules are, respectively,

(a) $$\frac{50.0 \text{ g Ag}}{107.87 \text{ g/g-atm}} = 0.463 \text{ mole Ag}$$

(b) $$\frac{50.0 \text{ g Cu}^{2+}}{63.54 \text{ g/g-ion}} = 0.787 \text{ mole Cu}^{2+}$$

(c) $$\frac{50.0 \text{ g CO}_2}{44.01 \text{ g/g-molecule}} = 1.14 \text{ moles CO}_2$$

MEASUREMENT OF NUMBERS OF PARTICLES

1. *Avogadro's number by densities of crystals*

It is possible to combine a knowledge of the interatomic spacings of a crystalline solid with its density and the weight of its gram-formula, and obtain a quite reliable measurement of the Avogadro number. The scattering of a beam of x-rays by a salt crystal acting as a diffraction grating toward the beam has permitted determinations not only of the arrangements of ions in the crystal but also of their distances apart. The production of x-ray diffraction patterns by the diffraction grating works very much on the same principle as a prism in producing a spectrum of colors when visible light is passed through it. Indeed, the same spectral effects are obtained when visible light passes through or is reflected from any transparent surface, such as a sheet of glass upon which many extremely fine, equally spaced, parallel, opaque lines are ruled.

Such an artificial diffraction grating rarely serves, however, for the production of an x-ray spectrum — not one, to be sure, of component wavelength colors, as with visible light. The wavelengths of x rays are far too short, as compared to visible light, to be diffracted by such grating lines. Rather, the spectrum being sought is one of wavelengths that are recordable upon a photographic plate. A crystal, on the other hand, serves as a natural medium for separation of x-radiation components because the far smaller interspacings and much more orderly arrangements of the particles in a crystal fulfil requirements for the diffraction.

Using a crystal of sodium chloride, NaCl, we illustrate how the distances between adjacent ions are coupled with density and with formula weight of the salt in establishing the value of N (Avogadro number) at 6.02×10^{23}. X-ray diffraction tells us that NaCl is a cubic lattice structure and that the centers of the Na^+ and Cl^- ions at adjacent lattice points are separated by an average distance of 2.819×10^{-8} centimeter. With the density of NaCl established at 2.165 g/c cr, and its mole (gram-formula) weight equal to 58.443 g, the following transpositions may be made to the desired *mole* quantities representative of the Avogadro number:

$$\text{volume of 1 mole NaCl} = \frac{\text{mass of 1 mole of NaCl}}{\text{density of NaCl}} = \frac{58.443 \text{ g/mole}}{2.165 \text{ g/cm}^3}$$

$$= 26.999 \text{ cm}^3/\text{mole}.$$

As the volume of a cube is the length of its edge cubed, it follows that the length of any side of a cube that is 26.999 cm³ in volume must be

$$\text{edge length of NaCl mole-cube} = \sqrt[3]{26.999 \text{ cm}^3}$$

$$= 3.000 \text{ cm/edge of NaCl mole-cube}.$$

Inasmuch as studies of diffraction have placed adjacent ions 2.819×10^{-8} centimeters apart, it follows that for every interval of 2.819×10^{-8} centimeters there is found one ion (irrespective of identity, Na^+ or Cl^-). Consequently,

$$\text{number of ions at edge} = \frac{3.000 \text{ cm}}{2.819 \times 10^{-8} \text{ cm}}$$

$$= 1.064 \times 10^8 \text{ ions}.$$

Therefore, for the entire cube containing the mole of NaCl (considered as a gram-formula unit),

$$\text{total number of ions} = (1.064 \times 10^8)^3$$

$$= 1.205 \times 10^{24} \text{ ions/mole}.$$

But, of the total number of ions in the cube of NaCl, half are Na^+; the other half, Cl^-. That is, a mole of NaCl, considered as a formula unit, actually contains 1 mole of independent Na^+ ions and 1 mole of independent Cl^- ions, subject to the electrostatic stresses of each other's opposite charges. Therefore,

number of Na^+ ions/mole NaCl

$$= \text{number of } Cl^- \text{ ions/mole NaCl}$$

$$= \frac{\text{total numbers of } Na^+ \text{ and } Cl^- \text{ ions/mole NaCl}}{2}$$

$$= \frac{1.205 \times 10^{24}}{2}$$

$$= 6.02 \times 10^{23} \text{ ions of } Na^+ \text{ or of } Cl^-/\text{mole NaCl}.$$

This, then, is N — the Avogadro number.

   2. *Avogadro's number by alpha-decay*

cation other than H^+ and an anion other than OH^-. In conformity with these definitions, salts were correspondingly interpreted as the chemical products formed, along with H_2O, when the H^+ of acids *neutralized* the OH^- of bases.

DISSOCIATION VERSUS IONIZATION

Before proceeding, it might be well to distinguish between the terms *dissociation* and *ionization*, inasmuch as the latter term is a somewhat more apt description of the aspects of the Arrhenius delineations that still remain valid. Admittedly, however, the rather loose interchangeability of these terms, as frequently encountered, has made differentiation between them somewhat less than critical. Nonetheless, a semblance of modern distinction will be maintained here as follows:

Ionization (as a process): This connotes the formation of ions by partial or complete cleavage or splitting of a polar chemical substance as the result of the breaking of its normal covalent bonds. This is the case when a polar solute such as $HC_2H_3O_3$ (acetic acid) is dissolved in a polar solvent such as water, reaction occurring in this instance via $HC_2H_3O_2 + H_2O \rightleftarrows H_3O^+ + C_2H_3O_2^-$.

Dissociation (as a process): This connotes the imparting of mobility or freedom of movement to ions already initially present as such but restricted in their chemical manifestations because of their mutual interionic electrostatic attractions. Thus, in conformity with present-day theories of the formation of complexions, the *ligands* — molecules or ions coulombically associated in a common coordination sphere — dissociate rather than ionize. Illustratively, the complex ions, $Cu(NH_3)_4^{2+}$ and $Cd(CN)_4^{2-}$, dissociate into ions of Cu^{2+} plus molecules of NH_3, and into ions of Cd^{2+} plus ions of CN^-, respectively. Likewise, solid NaCl — a completely ionic compound (Na^+Cl^-) — when dissolved in water, dissociate into Na^+ and Cl^- ions, rather than ionize. Gaseous HCl, however, a compound already predominantly nonionic in its initial state, ionizes (rather than dissociates) into H_3O^+ and Cl^- ions when it is dissolved in water.

WEAK VERSUS STRONG ARRHENIUS ELECTROLYTES

The magnitudes of ionization of different electrolytes are determined experimentally not only by direct measurements of electrical conductivity but also from the effects they produce upon the physical constants of the pure solvent. These effects of a solute include the abnormal elevation of the boiling point of the solvent, depression of its freezing point, lowering of its vapor pressure, and increase of its osmotic pressure. All these so-called *colligative properties* of solutions are accountable to the increases in the total numbers of particles present in the solution of an ionizing or dissociating solute.

Moreover, the increases in the equivalent conductances of many solutions as they are progressively diluted, and which culminate asymptotically in an experimental maximum (theoretically reconcilable with infinite dilution limits beyond which further additions of solvent yield no further increases in conductance) is certainly a convincing argument for the validity of partial dissociations of many electrolytes. Its unqualified categorical application today to weak acids, weak bases, and certain highly covalent "salts" (perhaps best called *pseudo-salts*) is solidly established and pedagogically consistent with all experimental revelations.

The classical applications of the theory of "partial" ionization to strong electrolytes leads not only to experimental frustration but also to the contradiction of experimental evidence that most solid-state substances classified as electrolytes yielding highly conducting water solutions when solubilized are already completely ionic in their lattice structures. This is validly revealed by x-ray "spot patterns" obtained by techniques initiated by W. H. and W. L. Bragg (England, 1913).

There still remains, however, the necessity of reconciling the theoretical 100% ionic character of so many electrovalent compounds with the failure of experiments to achieve 100% of electrical conductance, or corresponding measurements of the colligative properties (freezing point, boiling point, vapor pressure, osmotic pressure) of the respective solutions. This is satisfactorily resolved by considerations of interionic attractions and resultant "drag effects" that electrostatically restrict the independence and mobility of ions when, as in solutions of strong electrolytes, they are present in high concentrations.

This concept, primarily that of Debye and Hückel (1923), thus plausibly corrects the Arrhenius theory that strong electrolytes, as well as weak, undergo a reversible ionization and ion-association process (that is, $NaCl \rightleftarrows Na^+ + Cl^-$). These theories of "electrical drag" postulate that in solutions of strong electrolytes no actual discrete molecular species are either present or formed to account for the failure of experiments to obtain a 100% manifestation of ions in the aqueous solution of an exclusively electrovalent solute. Rather, it is to be considered that the ions are subject to electrostatically-formed ionic atmospheres and ion-pairing; that is, positive ions surrounded by negative ions, and negative ions, by positive ions. The effects may then be marked enough — with the relatively large numbers of ions deposited by strong electrolytes even in moderately dilute solutions, and increasingly so with progressively more concentrated solutions — to alter significantly the revelations of their total presence. Hence, a distinction is to be made between *actual* concentration and *apparent* concentration — the latter being the effective concentration manifested under experimental conditions.

The qualitative character of the picture here for strong electrolytes, which thus resolves the seeming deficiency of species known to be present but not manifested in experiments should, consequently, be formulated not by the nonexistent molecules of strong electrolyte (AB, CD_2, etc. — generalized) of Arrhenius' concept but, rather, by the ionic aggregates $[A^+B^-]$ $[D^-C^{2+}D^-]$etc., which indicate the proximity of mutual electrostatic hindrances to, or impairments of, the independence and free mobility of each ion. These ionic associations are called, variously, "bound ions," "ion pairs," or "ion clusters." Whatever their descriptive characterization, the usual matter of degree applies inasmuch as ions of different chemical identity, mass, and charge are subject to different electrical "drag effects." In accurate calculations, particularly those dealing with equivalent conductances, corrective computations with respect to these variances of free or unbound ions are often quite necessary.

The original Arrhenius theory provided plausible acceptable explanations of many experimental phenomena of the day and subsequently, not the least of which were the migrations of chemical species to the electrodes under the influence of a direct current of electricity, and of their ensuing chemical changes (*electrolysis*). These effects are consistent with the existence of positive and negative ions in solutions, and of the attracting of these ions to the respective oppositely charged electrodes. Too, the physical significance of an electrically dipole solvent like H_2O in promoting, by its attractions for susceptible polar molecules of solute or complex ions, increased cleavage of solute with progressive dilutions, is experimentally verified not only by measurements of equivalent conductances of the respective solutions but also by differences in chemical behavior. Thus, although ionized or dissociated chlorine-containing solutes may yield precipitations of silver chloride when treated with silver nitrate solution, virtually non-ionized substances like carbon tetrachloride (CCl_4) or chloroform ($CHCl_3$) do not yield the chloride ion required for electrostatic attraction to the silver ion and, consequently, no resultant precipitation of insoluble silver chloride.

The total Arrhenius theory even postulated stepwise dissociations of *polybasic* acids, which are today firmly established. Its failure, however, to delineate likewise as acids the electrically-charged hydrogen-bonded stepwise intermediates derived from parent acids (for example, HCO_3^- from H_2CO_3; HS^- from H_2S; $H_2PO_4^-$ and HPO_4^{2-} from H_3PO_4) represented an additional shortcoming of its total acid-base concept. Its further failure to provide a satisfactory exposition of, or even a suggestive emphasis upon, the contributory role of the solvent in the ionization process led, inevitably, to the present reappraisals of acid-base theory expounded in the Brönsted-Lowry concepts.

BRÖNSTED-LOWRY ACIDS AND BASES

In the modern concepts of ionization of aqueous solutes, the visualizations are:

1) an acid is any species, molecular or ionic, that can yield a proton (H^+ ion) in response to chemical demands made upon it by a base in aqueous environment.

2) a base is any species, molecular or ionic, that can accept a proton from an acid in aqueous environment.

These definitions bring into sharp focus the mutual functional relationships, lacking in the Arrhenius concept, of acid and base inevitably present together in an environment (a solution) conducive to their simultaneous chemical changes. The capacity of the acid to function as a proton donor is necessarily conditioned by the availability of a base to act as a proton acceptor; and vice-versa. The role of the solvent in the chemistry of the process of ionization or dissociation is now of equal significance with that of the solute, and this puts into logical perspective the contributions of the very many species that by Arrhenius' concepts would be classified as neither acid nor base despite their experimentally confirmed contributions to the over-all acidity or basicity of their solutions.

The descriptive relationships to be delineated may, perhaps, be best understood by first inspecting the following specific illustrations of Brönsted-Lowry formulation:*

a) $HCl + H_2O \rightleftarrows H_3O^+ + Cl^-$

b) $H_2SO_4 + H_2O \rightleftarrows H_3O^+ + HSO_4^-$

c) $HC_2H_3O_2 + H_2O \rightleftarrows H_3O^+ + C_2H_3O_2^-$

d) $Al(H_2O)_3(OH)_3 + H_2O \rightleftarrows H_3O^+ + Al(H_2O)_2(OH)_4^-$

e) $H_3O^+ + H_2O \rightleftarrows H_3O^+ + H_2O$

f) $NH_4^+ + H_2O \rightleftarrows H_3O^+ + NH_3$

g) $Al(H_2O)_6^{3+} + H_2O \rightleftarrows H_3O^+ + Al(H_2O)_5OH^{2+}$

h) $HFe(CN)_6^{2-} + H_2O \rightleftarrows H_3O^+ + Fe(CN)_6^{3-}$

i) $HPO_4^{2-} + H_2O \rightleftarrows H_3O^+ + PO_4^{3-}$

j) $(H_2N)_2CNH + H_2O \rightleftarrows C(NH_2)_3^+ + OH^-$

k) $Cr(H_2O)_3(OH)_3 + H_2O \rightleftarrows Cr(H_2O)_4(OH)_2^+ + OH^-$

l) $NH_3 + H_2O \rightleftarrows NH_4^+ + OH^-$

m) $Ba(OH)^+ + H_2O \rightleftarrows Ba(H_2O)^{2+} + OH^-$

n) $Cr(H_2O)_5OH^{2+} + H_2O \rightleftarrows Cr(H_2O)_6^{3+} + OH^-$

o) $NH_2^- + H_2O \rightleftarrows NH_3 + OH^-$

p) $H_2AsO_4^- + H_2O \rightleftarrows H_3AsO_4 + OH^-$.

---

*Arrow-shafts of different lengths are used to indicate significantly different over-all driving forces for acidity or basicity to left or right of the reversible reaction, when such qualitative emphasis is desired. Ordinarily, however, such refinements of the written equation are neither necessary nor convenient; or even possible unless the specific strengths of the acids and bases contributing to the over-all equilibrium constant of the reaction are known.

The following descriptions apply to the species of solute formulated as the first item in each of the foregoing equations:

a) hydrochloric acid, a strong molecular acid
b) sulfuric acid, a strong molecular acid
c) acetic acid, a weak molecular acid
d) hydrated aluminum hydroxide, a weak molecular acid
e) hydronium ion, a strong cationic acid
f) ammonium ion, a weak cationic acid
g) hydrated aluminum ion, a weak cationic acid
h) hexacyanomonohydrogen ferrate(III) ion, a strong anionic acid
i) monohydrogen phosphate ion, a weak anionic acid
j) guanidine, a strong molecular base
k) hydrated chromic hydroxide, a weak molecular base
l) ammonia, a weak molecular base
m) hydroxo barium ion, a strong cationic base
n) monohydroxo hydrated chromic ion, a weak cationic base
o) amide ion, a strong anionic base
p) dihydrogen orthoarsenate ion, a weak anionic base.

It will be noted that the Brönsted-Lowry concept of an acid as a proton donor does not alter the Arrhenius definition of an acid; but it does supplement it by including in the "acid" category ionic as well as molecular species. It also corrects the formerly molecular representations of many substances now known to be purely ionic — this, in conformity with x-ray diffraction patterns of their solid-state crystal structures. It is the concept of the base perhaps, that has required the more radical of the re-interpretations and revisions of the Arrhenius delineations, because many species of solutes without the hydroxyl radical impart markedly basic characteristics to their solutions.

CONJUGATE ACIDS AND BASES — RIVAL SYSTEMS

A careful perusal of the equations presented in the preceding section reveals a common theme — the competitive tendency among chemical species in aqueous media to seize or to retain a proton. Expressed in the opposite sense, and equally valid, is their tendency to divest themselves of, or remain independent of, proton associations — frequently only partially fulfilled, as in stepwise ionizations. How effectively each species performs in this contest will determine not only the role that it plays, as a proton donor — an acid — or a proton acceptor — a base — but, also the magnitude of such role and of the consequent contribution made to the over-all acidity or basicity of its aqueous environment.

It is seen that proton transference (*protolysis*) is represented in both forward and reverse processes of the reaction equilibria given. In other words, an acid-base reaction occurs not only between the added solute and the solvent but also between the chemical products formed. Hence, there exists in H_2O solutions of a simple solute — that is, a solute that provides in aqueous solubilization just two constituent species — two sets of acids and two sets of bases, each acid competing with the other in the divesting of a single proton, and each base likewise competing with the other in enforced acceptance of that single proton. This may be empirically described by the generalized equation:

$$acid_1 + base_2 \rightleftarrows acid_2 + base_1$$

in which acid and base represent the conjugated related cleavage species either of solute or of solvent; and $acid_2$ and $base_2$, the conjugates of the other. The chemical identity of each partner in its respective acid-base pair or combination is thus defined in terms of a deficient equilibrium or superfluity of a single proton; that is,

$$conjugate\ acid \rightleftarrows proton + conjugate\ base.$$

Just how far to left or to right a particular reaction proceeds as written in equation form, and the magnitude of its calculated equilibrium constant, are matters that depend upon the relative strengths of the competing acids and bases and these are, by no means, necessarily the same. It is functionally inherent in the formation of an acid that the weaker it is, the stronger is its conjugate base; and that the stronger the acid, the weaker must be its conjugate base.

AMPHOTERISM — BRÖNSTED-LOWRY CONCEPTS

Returning to the chemical equations previously given, it is important to observe the *amphoteric* character of the H_2O molecules; that is, their dual capacities to function both as acids and as bases. They respond as bases when in the presence of a substance with greater acidic or weaker basic strength than themselves — hence, their enforced acceptance of protons; and as acids when in the presence of a substance with greater basic or weaker acidic strength than themselves — hence, their divestment of protons. That amphoteric (or ampholytic) behavior is not restricted to the species of the solvent is readily apparent by inspection of the following equations which reveal the dual character of NH_3 — first, as the conjugate base of NH_4^+ (ammonium ion), and second, as the conjugate acid of NH_2^- (amide ion):

$$\left\{ \underset{base_1}{NH_3} + \underset{acid_2}{H_2O} \rightleftarrows \underset{acid_1}{NH_4^+} + \underset{base_2}{OH^-} \right\}$$

$$\left\{ \underset{base_1}{NH_2^-} + \underset{acid_2}{H_2O} \rightleftarrows \underset{acid_1}{NH_3} + \underset{base_2}{OH^-} \right\}$$

The second equation is noteworthy in its revelation that even a substance so completely familiar as a base as NH_3 has no proprietary right to function as such. Rather, the defining of its precise role, as well as the magnitude thereof, depends upon the presence or absence of stronger bases. In this instance, it is qualitatively depicted by the magnitude (size of arrow) of the forward reaction, which shows the displacement of the weaker base (NH_3) by the stronger one (NH_2^-). The relative extent of the reverse reaction in the first equation similarly conveys the fact that OH^- is a stronger base than NH_3.

The generalization may be made that strong bases displace weaker bases from solutions of salts, and that strong acids likewise displace weaker acids from solutions of salts. Thus, the addition of hydrochloric acid to a solution of sodium produces the considerably weaker acetic acid. It would be prudent, perhaps, to stress that a generalization such as this may be applied with confidence only to simple or uncomplexed acids or bases; that is, in the absence of strong bonding forces, covalent or electrostatic. Thus, illustratively, the strong base, NaOH, does not readily displace the ligand NH_3 molecules from the complex ions $Co(NH_3)_6^{3+}$, $Co(NH_3)_5Cl^{2+}$, and $Co(NH_3)_4Cl_2^+$.

Amphoteric or ampholytic behavior of species is always exhibited in the stepwise ionizations of polybasic (polyprotic) weak acids. Illustratively, the primary, secondary, and tertiary ionizations of phosphoric acid, H_3PO_4, are, respectively,

$$acid \; H_3PO_4 + H_2O \rightleftharpoons H_3O^+ + H_2PO_4^- \; base$$
$$acid \; H_2PO_4^- + H_2O \rightleftharpoons H_3O^+ + HPO_4^{2-} \; base$$
$$acid \; HPO_4^{2-} + H_2O \rightleftharpoons H_3O^+ + PO_4^{3-} \quad base.$$

Each successive ionization is weaker than the preceding, attesting to the progressively increasing strengths of the conjugate bases that are formed and the consequent increased extents of reversal of the respective ionizations. As observed, $H_2PO_4^-$ (the conjugate base of H_3PO_4 in the primary; functions ampholytically in the secondary as the conjugate acid of HPO_4^{2-}) and HPO_4^{2-} (the conjugate base of $H_2PO_4^-$ in the secondary) functions in the tertiary as the conjugate acid of PO_4^{3-}. Of all the species of solute present, only H_3PO_4 and PO_4^{3-} display no amphoterism. The former can act solely as an acid in aqueous medium, the latter, exclusively as a base.

Perhaps the elusiveness of truly complete success in achieving a definition of amphoterism that comprehensively encompasses, at one and the same time, all the facets and magnitudes of its essential acid-base mechanisms suggests itself in these polyprotic ionizations. Although on previous occasions the dual roles of an ampholyte were separately delineated in different reaction environments (acidic when the ampholyte functioned as a base, basic when the ampholyte behaved as an acid), here its dual capacity is being revealed simultaneously as both acid and base under precisely identical conditions of chemical environment. This is shown by the presence of $H_2PO_4^-$ in the concurrent equilibria of primary and secondary ionizations, and also by the presence of HPO_4^{2-} in equilibria of both secondary and tertiary ionizations. Perhaps, in the interests of clarity, a *simultaneous ampholyte* may be designated as a species capable of simultaneously giving up and accepting a proton in equilibrium responses to the demands of other species present. The preponderance of its relative reversible equilibrium balance is determined, by the specific acidic and basic strengths of all competing species, as well as of its own. There is an additional term that is occasionally applied to an amphoteric substance; namely, "amphiprotic." This term is generally interpreted in the specific sense of a simultaneous ampholyte. Clearly, and as indicated earlier, the solvent H_2O molecules are themselves amphiprotic substances engaged in simultaneous gain and loss of H^+, as represented by its self-ionization. Thus, the equation

$$H_2O + H_2O \rightleftarrows H_3O^+ + OH^-$$
$$\text{\small acid}_1 \quad \text{\small base}_2 \quad \text{\small acid}_2 \quad \text{\small base}_1$$

shows water molecules both giving and taking a proton under identical conditions.

PROTOLYSIS IN NONAQUEOUS SYSTEMS — ANHYDROUS ACIDS AND BASES

The phenomenon of amphiprotic behavior of a solvent by self-ionization is by no means limited to H_2O, but is seen as well with a number of nonaqueous polar solvents; for example, in the following equations using anhydrous acetic acid and pure liquid ammonia, respectively:

$$\left\{ HC_2H_3O_2 + HC_2H_3O_2 \rightleftarrows H_2C_2H_3O_2^+ + C_2H_3O_2^- \right\}$$
$$\text{\small acid}_1 \qquad \text{\small base}_2 \qquad \text{\small acid}_2 \qquad \text{\small base}_1$$

$$\left\{ NH_3 + NH_3 \rightleftarrows HN_4^+ + NH_2^- \right\}.$$
$$\text{\small acid}_1 \quad \text{\small base}_2 \quad \text{\small acid}_2 \quad \text{\small base}_1$$

This extension of the Brönsted-Lowry concepts to nonaqueous solvents, without conflicts, not only expresses a logical unity of theoretical approach but also affords valuable experimental opportunities for investigating the relative strengths of strong acids. These include perchloric acid ($HClO_4$), nitric acid (HNO_3), sulfuric acid (H_2SO_4, in its primary ionization), and the strongly water-ionized covalent hydrogen halides — HI, HBr, and HCl. All of these substances exhibit about the same acidic strength when dissolved in water, conforming to the greater basic strength of the H_2O molecule compared with that of the individual ions, ClO_3^-, NO_3^-, HSO_4^-, I^-, Br^-, and Cl^-, respectively; and thus to seize preferentially all available hydrogen ions through an enforced complete ionization of these strong acids. As formulated,

$$\left.\begin{array}{l} HClO_4 \\ HNO_3 \\ H_2SO_4 \\ HI \\ HBr \\ HCl \end{array}\right\} + \ H_2O \ \rightarrow \ H_3O^+ \ + \left\{\begin{array}{l} ClO_4^- \\ NO_3^- \\ HSO_4^- \\ I^- \\ Br^- \\ Cl^- \end{array}\right.$$

conjugate acid₁ (stronger than acid₂)   conjugate base₂ (stronger than base₁)   conjugate acid₂ (weaker than acid₁)   conjugate base₁ (weaker than base₂)

LEVELING EFFECTS — AQUEOUS AND NONAQUEOUS MEDIA

The total net effect of the behavior just delineated, whereby all the strong acids depicted "level" down to practically equal strength in water solution is a direct consequence of the greater affinity of the H_2O molecule for H^+ than for the respective competitive anionic conjugates of these strong acids. This must be interpreted as an inability of any water solution to retain significant concentrations of acids stronger than the hydronium ion, H_3O^+. It follows that if H_2O molecules, by virtue of their relative basic strengths, are able to level all strong acids to virtually identical strengths, the substitution of a more basic solvent species than H_2O molecules will virtually guarantee the maximum leveling effect. The very strongest acid that can possibly be retained by the *nonaqueous* pure liquid NH_3 (referred to previously) will be NH_4^+ which logically portends that even weak acids may become extensively, even fully, ionized in such a medium. This is corroborated. The weak acetic acid ($HC_2H_3O_2$), when dissolved in the pure liquid NH_3, becomes just about as strong as any of the strong acids in aqueous solutions that have already been mentioned. In accordance with the formulation of this reaction equilibrium between anhydrous $HC_2H_3O_2$ and NH_3:

$$HC_2H_3O_2 + NH_3 \rightarrow NH_4^+ + C_2H_3O_2^-$$

the leveling-effect displacement to the right is virtually complete.

In order to evaluate the relative strengths of strong acids it is clearly necessary to employ a solvent that is much less basic than H_2O, that is, one sufficiently acidic to circumvent any leveling effect. Such a solvent is anhydrous (nonaqueous) liquid acetic acid. Attainment of reversible equilibria in such an environment occurs short of the 100% protolytic transfers enforced by a water environment and provides a comparative distribution range of measured ionizations capable of differentiating among otherwise equally aqueous-strong acids solvation[1] and interionic attractions produc-

---

[1]Solvation: A general term designating the electrostatic attractions of polar solvent molecules to ions; compare with "hydration" when referring specifically to the solvent water.

tive of deviations do exist, and their effects may be calculated and compensated by appropriate mathematical corrections.

Familiar monoprotic strong acids in anhydrous solvent acetic acid give the following relative strength responses:

$$HCl < HNO_3 < HBr < HI < HClO_4$$
$$\longrightarrow \text{ increasing acidity } \longrightarrow$$

Formulated in reaction terms consistent with their respective displacements of the forward reaction in the attained reversible equilibria

$$
\left.\begin{array}{l} HCl \\ HNO_3 \\ HBr \\ HI \\ HClO_4 \end{array}\right\} + HC_2H_3O_2 \underset{\text{system}}{\overset{\text{anhydrous}}{\rightleftarrows}} H_2C_2H_3O_2^+ + \left\{\begin{array}{l} Cl^- \\ NO_3^- \\ Br^- \\ I^- \\ ClO_4^- \end{array}\right.
$$

$$\qquad acid_1 \qquad\qquad base_2 \qquad\qquad\qquad acid_2 \qquad\quad base_1$$

Again, it is appropriate to stress the purely relative meanings of the terms "acid" and "base," for here we find the familiar acetic acid functioning instead as a base. By the same token, the highly arbitrary categorical classifications of acids and bases as "strong" and "weak" can be given only expedient justification, and even then only when carefully considered in the light of all the competitive factors of a specific reaction environment. Obviously, many moderate variabilities exist between the extremes of weak and strong. As with strong acids, any attempt to measure the relative strengths of strong bases is likewise defeated in a system with water as a solvent, because now the water molecules act too acidically and compel, in effect, a leveling of all strong bases to a practically identical basicity — namely, that of the OH^-. As illustrated in the following formulation, OH^- remains the strongest base capable of significant existence in a water solution:

$$NH_2^- \quad + \quad H_2O \quad \rightarrow \quad NH_3 \quad + \quad OH^-$$

conjugate base$_1$	conjugate acid$_2$	conjugate acid$_1$	conjugate base$_2$
(stronger than base$_2$)	(stronger than acid$_1$)	(weaker than acid$_2$)	(weaker than base$_1$)

To measure the strength of bases stronger than H_2O requires the same means as in the measurement of strong acids — namely, the employment of a nonaqueous solvent — but, on this occasion one that is less acidic (that is, more basic) than H_2O and which, consequently, permits the circumvention of a leveling effect. Such a solvent is pure liquid ammonia.

Employment of nonaqueous solvents with significantly larger acidic strengths than H_2O (anhydrous H_2SO_4, for example) would not only level the many familiar aqueous-weak bases to the common categorical identity of strong bases but, as already noted, may well convert even aqueous-weak acids into respectable bases. Thus, note acetic acid functioning in the following as a nonaqueous base:

$$\underset{\text{base}_1}{HC_2H_3O_2} + \underset{\text{acid}_2}{H_2SO_4} \underset{\underset{\text{medium}}{\longleftarrow}}{\overset{\text{anhydrous}}{\longrightarrow}} \underset{\text{acid}_1}{H_2C_2H_3O_2^+} + \underset{\text{base}_2}{HSO_4^-}$$

THE ACID-BASE EQUATION — POSSIBLE NET DISAPPEARANCE OF H_2O

We have, so far, devoted attention rather exclusively to the equilibria of single-solute solutions with the objective of not only portraying but also stressing the chemical role of the solvent in its relationship to the solute as expressed by the Brönsted-Lowry theories. Chemical work, however, must contend inescapably with the many species of solute that are present simultaneously and which, in keeping with the differences in their electronic make-ups, are either stronger acids or stronger bases than H_2O. When multiple species of solute are concurrently present, H_2O is frequently both weaker acid and weaker base simultaneously than any of the species of solute present. The net effect of this relationship, as already stressed, is the formation from H_2O molecules of H_3O^+ by the stronger acid and of OH^- by the stronger base. In consequence of the resultant requirements for neutralization,

$$H_3O^+ + OH^- \rightleftarrows 2H_2O,$$

the possibility exists that H_2O may disappear as an entity from the net equation for a reaction because it is cancelled out arithmetically in equivalent amounts on both sides of the equation. This is illustrated in the following derivation describing the reaction between aqueous solutions of acetic acid and ammonia:

$$HC_2H_3O_2 + H_2O \rightleftarrows H_3O^+ + C_2H_3O_2^-$$
$$NH_3 + H_2O \rightleftarrows NH_4^+ + OH^-$$

adding $\quad HC_2H_3O_2 + NH_3 + 2H_2O \rightleftarrows NH_4^+ + C_2H_3O_2^- + \underbrace{H_3O^+ + OH^-}_{= 2H_2O}$

The water-cancelled "net" identifies itself directly with the Brönsted-Lowry formulation,

$$\underset{\text{acid}_1}{HC_2H_3O_2} + \underset{\text{base}_2}{NH_3} \rightleftarrows \underset{\text{acid}_2}{NH_4^+} + \underset{\text{base}_1}{C_2H_3O_2^-}$$

If OH^-, rather than NH_3, were the solute base in reaction with $HC_2H_3O_2$, the "net" would show H_2O as an uncancelled chemical product, thus:

$$\underset{\text{acid}_1}{HC_2H_3O_2} + \underset{\text{base}_2}{OH^-} \rightleftarrows \underset{\text{acid}_2}{H_2O} + \underset{\text{base}_1}{C_2H_3O_2^-}$$

The omission, then, of H_2O from the over-all picture of chemical change, when the net equation so invites, is not to be construed as a neglect of the principles expounded with respect to the chemical role of the solvent, but rather as an expression of expediency. Not only is it a matter of practical convenience to use a net equation but, indeed, it is also frequently necessary to do so when experimental certainty of the exact mechanism of stepwise reaction is lacking, even though plausible surmises may be made of the intermediates involved. It is common practice, therefore, deliberately to ignore the presence of H_2O (provided it is not an uncancelled reaction product) whenever the chemistry of the solute can be conveniently interpreted without it. This is of particular advantage in the many mathematical computations that involve reaction-induced changes of concentrations of the solute species present. With the invariably highly diluted solutions used in aqueous chemical reactions, the concentration of the solvent H_2O undergoes no significant reaction change, as a rule, in relation to the solute. Taking the density of pure water as 0.997 g/ml at customary laboratory temperatures (about 25°C) its concentration would be

$$\frac{997 \text{ g/liter}}{18.0 \text{ g/mole}} = 55.3 \text{ moles/liter.}$$

This value is already so large initially that a slight decrease or increase in percentage by normal reaction does not impair its assignment as a constant in conventional calculations. As such it is then conveniently combined with the equilibrium constant for the particular reaction.

It is frequently considered advantageous, as well, to neglect H_2O in identity-formulations of hydrated species, where "hydration" refers not to the loose H_2O molecules in the vicinity of a charged species, but rather to the coordinated (attached) molecules of dipole solvent that are part of the coordination sphere of an ion. Inasmuch as all charged species in aqueous solution are considered to be hydrated, anions as well as cations, and not merely the proton which hitherto alone has been given this distinction — that is, as $H(H_2O)^+$ or H_3O^+, hydronium ion — the exclusion of coordinated water of hydration becomes quite general.

This holds whenever electronic structural relationships of species are not the subject of exposition. The frequent preferential treatment familiarly accorded the hydrogen ion in transposing it to the hydrated version is largely a concession to the extremely high charge density (ratio of charge to size of ionic radius) of the bare H^+ which prevents it from existing as such in aqueous solution. Experimental evidence is firmly established with respect to this. An additional reason is the especial importance of this ion in its influence, in conformity with its pH upon the course and extent of nearly all aqueous chemical reactions. The quite general lack of experimental information concerning the degree of hydration (the hydration number) as concentration and temperature change during the course of a reaction is certainly ample justification for the simplifications that are expediently made and which often excusably characterize the hydrogen ion as unhydrated H^+.

As a matter of fact, the formulation H_3O^+ is purely an empirical representation of $H(H_2O)_x^+$ wherein, in response to variable conditions of temperature and concentration, the value of x may be 1, 2, 3, or 4. These simplifications are most welcome when offered in equations as alternately acceptable expressions of what otherwise may prove a burdensome task of writing, as illustrated in the following step-wise ionizations of the hydrated chromium(III) ion in reaction with OH^-:

$$\text{Primary } Cr(H_2O)_6^{3+} + OH^- \rightleftarrows Cr(H_2O)_5(OH)^{2+} + H_2O$$
$$\textit{alternate } Cr^{3+} + OH^- \rightleftarrows Cr(OH)^{2+}$$

$$\text{Secondary } Cr(H_2O)_5OH^{2+} + OH^- \rightleftarrows Cr(H_2O)_4(OH)_2^+ + H_2O$$
$$\textit{alternate } CrOH^{2+} + OH^- \rightleftarrows Cr(OH)_2^+$$

$$\text{Tertiary } Cr(H_2O)_4(OH)_2^+ + OH^- \rightleftarrows Cr(H_2O)_3(OH)_3 + H_2O$$
$$\textit{alternate } Cr(OH)_2^+ + OH^- \rightleftarrows Cr(OH)_3[\text{precipitated}].$$

We carry the point farther by showing the continuing aspect of amphoteric behavior that is being developed. Here the precipitated, molecularly neutral chromic hydroxide of tertiary formation, which is scarcely effectual as an ampholytic acid when responding to a base as weak as H_2O, functions most effectively as an acid in neutralizing further additions of the strong base OH^-. Its capacity to dissolve fully therein by solubilization is shown thus:

$$Cr(H_2O)_3(OH)_3 + OH^- \rightleftarrows Cr(H_2O)_2(OH)_4^- + H_2O$$
$$\textit{alternate } Cr(OH)_3 + OH^- \rightleftarrows Cr(OH)_4^-.$$

Thermal decomposition of the $Cr(OH)_4^-$ ion, once formed, is frequently of sufficient extent to warrant its being formulated as CrO_2^-, dehydration occurring as follows:

$$Cr(OH)_4^- \rightleftarrows CrO_2^- + 2H_2O.$$

This behavior is quite general with hydroxy-coordinated complexes, but inasmuch as it is difficult to evaluate experimentally the extent of dehydration as a function of the temperature variant, the choice of formula is hardly critical. This behavior is cited primarily to prepare the student, who may be using this text in conjunction with other readings, for the usual diversities that will be encountered in descriptive terminology — even in definitions, when specific reaction-media conditions (temperature, concentration, pH, etc.) are not stated as a common reference point.

The ampholytic functioning of the precipitated $Cr(OH)_3$ likewise accounts for its complete dissolution in acids of sufficient strength in which the hydroxide functions as a neutralizing base. Thus, the neutralization of $Cr(OH)_3$, itself, by H_3O^+ would, in fully simplified form, proceed successively as follows:

$$Cr(OH)_3 + H^+ \rightleftarrows Cr(OH)_2^+ + H_2O$$
$$Cr(OH)_2^+ + H^+ \rightleftarrows Cr(OH)^{2+} + H_2O$$
$$Cr(OH)^{2+} + H^+ \rightleftarrows Cr^{3+} + H_2O$$

with a "net" of

$$Cr(OH)_3 + 3H^+ \rightleftarrows Cr^{3+} + 3H_2O.$$

It is evident that in order for a molecularly neutral substance to function amphoterically it must be sufficiently acidic to yield anions in basic media, and sufficiently basic to yield cations in acidic media.

RELATIVE STRENGTHS OF SOME ACIDS AND BASES

Table 2:1, showing various aqueous acid-base conjugates, provides a convenient qualitative appraisal of their relative strengths. In some instances the relative strengths are so very close that rearrangements of their respective orders may well occur under the variabilities of experimental working conditions, and, perhaps, even dictated when present measurements are further refined.

TABLE 2:1

$$\text{Acid} \xleftarrow[\text{conjugates}]{\text{protolysis}} \text{Base}$$

$$\frac{[H_3O^+][\text{Base}]}{[\text{Acid}][H_2O]} = K \text{ equil.} \qquad\qquad K \text{ equil.} = \frac{[\text{Acid}][OH^-]}{[H_2O][\text{Base}]}$$

	Strength *decreases descending*	*Strength* *increases descending*
aqueously leveled to H_3O^+	Perchloric Acid, $HClO_4$	ClO_4^-, Perchlorate Ion
	Hydriodic acid, HI	I^-, Iodide Ion
	Hydrobromic Acid, HBr	Br^-, Bromide Ion
	Nitric Acid, HNO_3	NO_3^-, Nitrate Ion
	Hydrochloric Acid, HCl	Cl^-, Chloride Ion
	Permanganic Acid, $HMnO_4$	MnO_4^-, Permanganate Ion
	Sulfuric Acid, H_2SO_4	HSO_4^-, Hydrogen Sulfate Ion
	Hydronium Ion, H_3O^+	H_2O, Water
	Iodic Acid, HIO_3	IO_3^-, Iodate Ion
	Sulfamic Acid, HNH_2SO_3	$NH_2SO_3^-$, Sulfamate Ion
	Oxalic Acid, $H_2C_2O_4$	$HC_2O_4^-$, Hydrogen Oxalate Ion
	Hydrogen Sulfate Ion, HSO_4^-	SO_4^{2-}, Sulfate Ion
	Chlorous Acid, $HClO_2$	ClO_2^-, Chlorite Ion
	Sulfurous Acid, H_2SO_3	HSO_3^-, Hydrogen Sulfite Ion
	Phosphoric Acid, H_3PO_4	$H_2PO_4^-$, Dihydrogen Phosphate Ion

TABLE 2:1 (Continued)

Iron (III) Ion, $Fe(H_2O)_6^{2+}$	$Fe(H_2O)_5(OH)^{2+}$, Monohydroxy Iron (III) Ion
Hydrofluoric Acid, HF	F^-, Fluoride Ion
Nitrous Acid, HNO_2	NO_2^-, Nitrite Ion
Arsenic Acid, H_3AsO_4	$H_2AsO_4^-$, Dihydrogen Arsenate Ion
Chromium (III) Ion, $Cr(H_2O)_6^{3+}$	$Cr(H_2O)_5(OH)^{2+}$, Monohydroxy Chromium (III) Ion
Hydrogen Oxalate Ion, $HC_2O_4^-$	$C_2O_4^{2-}$, Oxalate Ion
Acetic Acid, $HC_2H_3O_2$	$C_2H_3O_2^-$, Acetate Ion
Aluminum (III) Ion, $Al(H_2O)_6^{3+}$	$Al(H_2O)_5(OH)^{2+}$, Monohydroxy Aluminum (III) Ion
Carbonic Acid, H_2CO_3	HCO_3^-, Hydrogen Carbonate Ion
Hydrosulfuric Acid, H_2S	HS^-, Hydrogen Sulfide Ion
Dihydrogen Phosphate Ion, $H_2PO_4^-$	HPO_4^{2-}, Monohydrogen Phosphate Ion
Dihydrogen Arsenate Ion, $H_2AsO_4^-$	$HAsO_4^{2-}$, Monohydrogen Arsenate Ion
Hydrogen Sulfite Ion HSO_3^-	SO_3^{2-}, Sulfite Ion
Hypochlorous Acid, HClO	ClO^-, Hypochlorite Ion
Copper (II) Ion, $Cu(H_2O)_4^{2+}$	$Cu(H_2O)_3OH^+$, Hydroxy Copper (II) Ion
Hypobromous Acid, HBrO	BrO^-, Hypobromite Ion
Boric (ortho) Acid, H_3BO_3	$H_2BO_3^-$, Dihydrogen Orthoborate Ion
Ammonium Ion, NH_4^+	NH_3, Ammonia
Hydrocyanic Acid, HCN	CN^-, Cyanide Ion
Zinc (II) Ion, $Zn(H_2O)_4^{2+}$	$Zn(H_2O)_3OH^+$, Hydroxy Zinc (II) Ion
Hydrogen Carbonate Ion, HCO_3^-	CO_3^{2-}, Carbonate Ion
Hypoiodous Acid, HIO	IO^-, Hypoiodite Ion
Magnesium (II) Ion, $Mg(H_2O)_6^{2+}$	$Mg(H_2O)_5OH^+$, Hydroxy Magnesium Ion
Hydrogen Peroxide, H_2O_2	HO_2^-, Peroxy Hydrogen Ion
Monohydrogen Phosphate Ion, HPO_4^{2-}	PO_4^{3-}, Phosphate Ion
Monohydrogen Arsenate Ion, $HAsO_4^{2-}$	AsO_4^{3-}, Arsenate Ion
Hydrogen Sulfide Ion, HS^-	S^{2-}, Sulfide Ion
Water, H_2O	OH^-, Hydroxyl Ion
Ammonia, NH_3	NH_2^-, Amide Ion
Hydroxyl Ion, OH^-	O^{2-}, Oxide Ion

aqueously *leveled* to OH^-

HYDROLYSIS — CLEAVAGE OF WATER MOLECULES BY SOLUTE

Perhaps, at least superficially, the introduction here of the concept of *hydrolysis* may appear somewhat overdue. Actually, however, it is only the use of the term, not the concept itself, that has been delayed, because the theoretical development of hydrolysis has been quietly proceeding since Brönsted-Lowry delineations were introduced. Definitions of hydrolysis usually portray this type of acid-base reaction as one that occurs between a salt and water with resultant splitting of the H_2O molecules and consequent formation of associated (weakly ionized) species plus a concurrent, generally increased, amount of either free H_3O^+ or free OH^-. Thus, the acidic character of an ammonium chloride solution (pH < 7.0; turns blue litmus to red) is logically shown in the reaction:

$$\underset{acid_1}{NH_4^+} + \underset{base_2}{H_2O} \rightleftarrows \underset{base_1}{NH_3} + \underset{acid_2}{H_3O^+}$$

The nonappearance of Cl^- in this equation indicates its chemically unchanged role as a "spectator," unable, by reason of its extreme weakness as a base, to form an associated species. Both NH_4Cl and HCl in solution are 100% ionic and cannot exist therein as molecules. Similarly, the basic character of a sodium acetate solution (pH > 7.0; turns red litmus to blue) is plausibly expressed by the equation:

$$\underset{base_1}{C_2H_3O_2^-} + \underset{acid_2}{H_2O} \rightleftarrows \underset{acid_1}{HC_2H_3O_2} + \underset{base_2}{OH^-}$$

The noninvolvement of Na^+ in the reaction is, manifestly, an expression of its almost negligible behavior as an acid. It will form no associated material in the solution because $NaC_2H_3O_2$ and $NaOH$ are both 100% ionic and cannot exist therein as molecules. The rather rare occasions when hydrolytic action may occur and yet not yield a solution significantly different in acidity or basicity from that of the pure H_2O (neutral, pH $= 7.0$), are limited to salts containing ions that can associate simultaneously with both the H_3O^+ and the OH^- derived from the split molecules of water and form weakly ionized acids and bases of respectively equal strengths. Thus, although the "double-barreled" hydrolysis of ammonium acetate in H_2O is very extensive in its formation of associated molecular species, the pH of the resultant solution in equilibrium remains virtually at the neutral point of 7.0 pH. This is so because the respective strengths of the acid NH_4^+ and the base $C_2H_3O_2^-$ are about equal, and consequently the strengths of their conjugates are likewise nearly identical. This conforms to the following derivation:

$$
\begin{aligned}
NH_4^+ + H_2O &\rightleftarrows NH_3 + H_3O^+ \\
C_2H_3O_2^- + H_2O &\rightleftarrows HC_2H_3O_2 + OH^-
\end{aligned}
$$

$$\text{net} \quad \underset{acid_1}{NH_4^+} + \underset{base_2}{C_2H_3O_2^-} + 2H_2O \rightleftarrows \underset{base_1}{NH_3} + \underset{acid_2}{HC_2H_3O_2} + 2H_2O$$

equal strengths   equal strengths

An aqueous solution will likewise retain a neutral character when the solute offers no species whatever that is chemically able to split H_2O molecules and associate with its cleavage products (H_3O^+ or OH^-) to form a covalently-bonded species. Thus, the solubilizing of KNO_3 in H_2O yields no hydrolytic alteration in the pH of the solution. This is completely consistent with the negligible acidic behavior of K^+ as an acid, the likewise insignificant functioning of NO_3^- as a base, and the consequent inability of molecular KOH and HNO_3 to form.

It inevitably follows that with any given simple salt the greater the degree of disparity in relative strengths of its cation (functioning as an acid) and its anion (functioning as a base), the greater will be the hydrolytic alteration in the pH of the solution as compared to the solvent. The larger the extent of formation of associated conjugate base (cation hydrolysis) compared to conjugate acid (anion hydrolysis), the greater the amount of unneutralized H_3O^+ and the smaller the pH. The greater the comparative amount of associated conjugate acid, the greater the concentration of unneutralized OH^- and the larger the pH. The same considerations lead validly to comparisons of the relative pH values of different solutions. Thus, given the separate solutions of the salts (a) Na_3PO_4, (b) Na_2HPO_4, and (c) NaH_2PO_4, the respective hydrolysis reactions are as follows:

(a) $$PO_4^{3-} + H_2O \rightleftarrows HPO_4^{2-} + OH^-$$
(b) $$HPO_4^{2-} + H_2O \rightleftarrows H_2PO_4^- + OH^-$$
(c) $$H_2PO_4^- + H_2O \rightleftarrows H_3PO_4 + OH^-$$

Inasmuch as the HPO_4^{2-} (K ionization, circa 10^{-13}) produced in (a) is much more weakly ionized than the $H_2PO_4^-$ (K ionization, circa 10^{-8}) produced in (b), and this in turn is much more weakly ionized than the H_3PO_4 (K ionization, circa 10^{-3}) produced in (c), it follows that of the three salts the most extensively hydrolyzed, and consequently the most basic, will be the Na_3PO_4. Similarly, the extent of hydrolysis and resultant basicity of the Na_2HPO_4 solution likewise exceeds that of the NaH_2PO_4. By the same process of appraising ionization constants, generally valid qualitative interpretations of the extent of hydrolytic behavior and of relative pH of the solutions may be obtained with completely unrelated salts.

HYDROLYSIS VERSUS IONIZATION

With the fundamental connotations of inherited classical terms established, it would seem that chemists have gone out of their way in search of variations expressive of the same thing, namely, the acid-base reaction. This, inescapably, is the fundamental unifying principle involved in ionization, amphoterism, hydrolysis, and other concepts (such as formation of complex ions) shortly to be developed. It must be remembered that all these concepts were developed more or less independently during the long history of the chemistry of acids and bases, and that the perception of their common central theme had to await interpretative evolution. Nonetheless, the different developments do perform a useful pedagogical service in emphasizing the various individual aspects of the total concept. It is this approach that invites further exploration.

The ionization of the solvent, H_2O, is clearly inherent in the capacity of the cation of a salt to function in forming a weakly ionized base, with a concomitant increase of H^+ in solution and of its anion, to form a weakly ionized acid and increase the OH^- of the solution. We bear in mind the definition of ionization already made — a process restricted to cleavage of an associated species into ions (HF in $HF \rightleftarrows H^+ + F^-$) and not accretion into ions (NH_3 in $NH_3 + H^+ \rightleftarrows NH_4^+$) which would be characteristic of the chemical functioning of all species of whatever identity depicted in reversible equilibrium in ionic solutions leading to the formation of ion-conjugates. Let us observe the conformity of definition here as applied to certain older formulations still widely in use, because of either expediency or, perhaps, of force of habit.

As a specific case, consider the representation of ammonia, NH_3, in the classical Arrhenius formulation as NH_4OH; it is still frequently written in this fashion. Modern Brönsted-Lowry theory produces the following contrasts in concept of this base (as with other "ammine" derivatives):

$$\text{Brönsted-Lowry} \quad NH_3 + H_2O \rightleftarrows NH_4^+ + OH^-$$
$$\text{Arrhenius} \quad NH_4OH \rightleftarrows NH_4^+ + OH^-$$

The first equation certainly depicts no ionization of NH_3 in the sense in which ionization has been defined — that is, cleavage, not accretion. It is from the second reaction that the familiar reference to the ionization of NH_3 derives — a descriptive characterization that is firmly established despite the complete lack of experimental evidence for the existence of such a compound as NH_4OH.

The reverse side of the picture involving the hydrolysis of NH_4^+,

$$NH_4^+ + H_2O \rightleftarrows NH_3 + H_3O^+$$

is properly visualized in the definition already made of hydrolysis, which requires the ionization of the solvent H_2O molecules (again by cleavage and not by accretion) only by reviving once more the nonexistent Arrhenius base, NH_4OH. Thus, as formulated —

$$NH_4^+ + \quad H_2O \rightleftarrows NH_4OH + H^+$$
or $$NH_4^+ + 2H_2O \rightleftarrows NH_4OH + H_3O^+$$

These exercises of the variable prerogatives of formulation and definition ("definition" always being "what you make it") are cited not only to mark more clearly the paths of chemical vocabulary — which if not carefully recognized will surely lose themselves in a wilderness of generalities — but also to prepare the student for the diversities of approach normally encountered in supplementary readings. Parenthetically, it may appear that the reason for not specifically characterizing the reaction $NH_3 + H_2O \rightleftharpoons NH_4^+ + OH^-$ as "hydrolysis" is that NH_3 is not a salt (in conformity with the limited frame of reference to hydrolysis adopted with respect to electrolytic reactions).

It must be pointed out, however, that hydrolysis is indeed a legitimate concept extended to reactions of purely molecular solute as well as ionic solute with water, with or without the formation of ions. Thus, the highly familiar hydrolysis of the organic compound thioacetamide, CH_3CSNH_2, is utilized widely in laboratories for the convenience it offers in safely supplying needed H_2S for the sulfide precipitations that characterize efficient schemes of qualitative analysis for separating and detecting cation. As revelations of the net mechanism of such hydrolysis the following equations apply:

$$\underset{\textit{Thioacetamide}}{CH_3CSNH_2} + H_2O \rightleftharpoons \underset{\textit{Acetamide}}{CH_3CONH_2} + H_2S$$

which expresses a very slow reaction at room temperature in the absence of added acid or base; and

$$\underset{\textit{Thioacetamide}}{CH_3CSNH_2} + 2H_2O \rightleftharpoons \underset{\textit{Acetate ion}}{C_2H_3O_2^-} + NH_4^+ + H_2S$$

which expresses the very extensive reaction promoted at elevated temperatures and in the presence of added acid or base.

THE CATEGORY OF SALTS — THE IMPLICATIONS

A true salt is defined by Arrhenius as a substance that simultaneously supplies cations other than H^+ and anions other than OH^-, formed concurrently with H_2O by neutralization. Implicit, however, in Brönsted-Lowry development is the essential ionic character of a salt, not as a fundamental composite unit in itself (proved experimentally to be nonexistent) but, rather, as a partnership of chemically separate, independent non-H^+ acid and independent non-OH^- base. In the solid state these species of ions are physically associated together because of the mutual attractions of their opposite electrical charges. In aqueous solution, these ions dissociate from one another and function as individuals, although their activities may be impaired to varying extents depending on concentrations, and which may cause ion-pairing or aggregation.

This represents the fundamental difference between the "true salt" and the "pseudo salt." The latter, a typically covalent molecular substance of characteristically low melting and boiling points (as well as of negligible electrical conductivity in the molten state) is like the true ionic salt only in supplying the other-than-H^+ and other-than-OH^- acid and basic constituents when dissolved in and ionized by solvent H_2O. "Pseudo-salts," which are few in number, are encountered among halides of metals chiefly in $+3$ and $+4$ states of oxidation — for example, $SnCl_4$, $FeCl_3$, and $AlCl_3$ or Al_2Cl_6 — although other states of oxidation and acid derivatives are represented in the classification — for example, $HgCl_2$, $Pb(C_2H_3O_2)_2$, and $Hg(CN)_2$. As true salts are in effect merely physical associations of electrostatically-attracted separate and chemically distinct acids and bases, their categorical classification as composite units serves no significant purpose other than a certain convenience of reference. As a class, however, they provoke certain speculations with respect to the true categorical identities of a number of familiar compounds traditionally regarded as bases — specifically, the hydroxy compounds of the elements in Periodic Groups I and II. Nearly all of these are essentially 100% ionic by virtue of the very low ionization potentials of the metals of these Groups.

In the Brönsted-Lowry definition, compounds such as NaOH, KOH, $Ba(OH)_2$, $Ca(OH)_2$, etc., can hardly be regarded as bases in their composite written forms inasmuch as none of them functions completely as a proton acceptor. Rather, it is the OH^- ion that alone functions as a base in its acquisition of H^+, and the metallic ion must be regarded as an acid, even though it may be negligible as such. But, this is precisely the definitive nature of a salt — a dissociable aggregate of ionically free and independent acid and base. The obvious dilemma that is posed is one of reconciling the usage of the chemical vocabulary of modern theoretical concept with that still firmly entrenched in the classic Arrhenius tradition. Certainly, to declaim against the impropriety of reference to KOH and NaOH as bases rather than salts — for the latter are, inescapably, their Brönsted-Lowry character — would indeed "raise eyebrows," habit being what it is.

It should therefore be tacitly understood that KOH, NaOH, and all other molecularly formulated ionic hydroxides that cannot function entirely as independent units, are Arrhenius bases in the sense that they supply the base OH^-. The continued reference to them as bases, then, will be acceptable and non-critical procedure when so tacitly interpreted. The anomalies of nomenclature are directly attributable to the historical fact that hydroxides were the sole bases originally conceived, as all Arrhenius formulations reveal. Today it is indisputably recognized that, other than with respect to the strength of the base involved, OH^- functions no differently as a base in aqueous media than do $C_2H_3O_2^-$, S^{2-}, CN^-, PO_4^{3-}, CO_3^{2-}, or any one of many dozens of other familiar anions that when allied with highly electropositive cations represent the category of salts.

NEUTRALIZATION VERSUS HYDROLYSIS — THE pH OF SALT SOLUTIONS

One other item deserves some interpretative clarification — the process of aqueous neutralization in its relationship to hydrolysis. Equating the neutralization reaction between NaOH and HCl yields

$$Na^+ + OH^- + H_3O^+ + Cl^- \rightleftarrows 2H_2O + Na^+ + Cl^-$$

The sodium and chloride ions, it may be noted, play no part in the chemical reaction. They are purely "spectators." Only the H^+ and the OH^- ions respond to the chemical change inherent in the formation of the water molecule. Any reference, therefore, to the formation here of the salt, NaCl, must be qualified as purely a potentiality of recovering the solid salt by evaporating the solvent.

If both HCl and NaOH are introduced into H_2O in equivalent amounts the neutralization leaves a truly neutral solution — that is, one with a pH of 7.0 wherein pH is a common logarithmic designation of the molar concentration of the H^+ ion (particularly convenient when the latter is very small); $pH = -\log_{10}[H^+]$ or, as exponentially expressed, $[H^+] = 10^{-pH}$. Neutralization, however, as a mechanism involving the formation of weakly ionized species and ions of "spectator" salts, need not produce a neutral solution. Thus, adding the weak base NH_3 to the strong acid HCl in equivalent amounts — equated as

$$NH_3 + H_3O^+ + Cl^- \underset{\leftarrow \text{hydrolysis}}{\xrightarrow{\text{neutralization}\rightarrow}} H_2O + NH_4^+ + Cl^-,$$

and adding the weak acid $HC_2H_3O_2$ to the strong base NaOH in equivalent amounts — equated as

$$HC_2H_3O_2 + Na^+ + OH^- \underset{\leftarrow \text{hydrolysis}}{\xrightarrow{\text{neutralization}\rightarrow}} H_2O + Na^+ + C_2H_3O_2^-,$$

in each instance reveals mutual neutralization of H^+ and OH^-, but in neither case is a neutral solution obtained. In conformity with theories already considered, the hydrolysis of NH_4^+ (reversal stage of the first equation) is of sufficient extent so that the solution is definitely acidic (pH less than 7.0) at the *equivalence point* of this reaction; that is, the point at which equivalent amounts of acid and base have reacted in conformity with the stoichiometry of the forward reaction.

Likewise in the second reaction, the hydrolysis of the $C_2H_3O_2^-$ proceeds sufficiently to yield a definitely basic solution (pH greater than 7.0), as indicated by the reversal stage of the written equation. Thus, although the processes of neutralization and hydrolysis are in a sense opposites, by no means do they necessarily have equal negating effects with respect to a solution's attainment of neutral pH. If the ions of salt present upon neutralization are of equal hydrolyzing strengths, as occurs with a water solution of the salt, $NH_4C_2H_3O_2$ (ammonium acetate), the solution will indeed be neutral. Here the virtual neutrality of the solution is assured because the weak base, NH_3, and the weak acid, $HC_2H_3O_2$, concurrently being formed, are both ionized to practically equal extents. That is, the hydrolytic strength of NH_4^+, reacting as an acid with water to form H_3O^+, is equal to that of $C_2H_3O_2^-$, reacting as a base to produce an equivalent amount of OH^-.

LEWIS ACIDS AND BASES

The Brönsted-Lowry concepts of acid-base behavior are generally adequate and effective in interpreting the different aspects of nonredox reactions of aqueous ionic solutions, and, in fact, they represent for this text the *modus operandi* of its mathematical applications. Nonetheless, it is valid here to appraise the extent of integration of such aqueous ionic behavior within the total acid-base medium because they must necessarily serve also to interpret the behavior of reactions in nonaqueous media, including gaseous and molten systems. In nonaqueous systems no protolytic transfers may be possible for the simple reason that the reacting species may be completely devoid of hydrogen atoms or ions — unlike water media wherein their presence is always assured. The objective here, therefore, is to reveal the representative place of the ionic solution in the general scheme of things because, in the emphasis it places upon H^+, the contributions of the Brönsted-Lowry concept to a uniform "field" theory applicable to all manner and type of acid-base reaction is obviously limited. Thus, in the conversion of barium oxide, BaO, to the sulfate, $BaSO_4$ (both substances are completely ionic), there would certainly seem to be a common mode of association or union of the reacting species. This might be in aqueous medium by reaction with H_2SO_4 solution in accordance with the equation:

$$BaO + H^+ + HSO_4^- \xrightarrow{\text{aqueous}} BaSO_4 + H_2O$$

or in nonaqueous medium as accomplished by impregnating the solid BaO with gaseous sulfur trioxide as in equation:

$$BaO + SO_3 \xrightarrow{\text{nonaqueous}} BaSO_4.$$

An additional illustration would be the quite similar conversions of CaO to $CaCO_3$, as accomplished in aqueous medium by boiling CaO with Na_2CO_3 solution, as equated:

$$CaO + CO_3^{2-} + H_2O \xrightarrow{\text{aqueous}} CaCO_3 + 2OH^-$$

and in nonaqueous medium by impregnating the oxide with CO_2 gas, as equated:

$$CaO + CO_2 \xrightarrow{\text{nonaqueous}} CaCO_3.$$

The logical recognition of the implicit existence of a common structural buildup of the products formed in reactions such as those given was shown by G. N. Lewis (United States, 1923), and it emphasizes the fundamental electronic feature responsible for the essential unity of all net over-all acid-base alterations, regardless of the chemical environments in which they occur. The Lewis theory entails the following definitions:

1. A base is any species — ionic, atomic, or molecular — that has a pair of electrons, as yet unshared, which it can supply for coordinate covalent attachment with another species; and

2. An acid is any substance — again, ionic, atomic, or molecular — that needs a pair of electrons and therefore seeks coordinate covalent attachment.

In simplified conformity with the octet theory of atomic structure and the stabilities assigned to the periodic configurations of inert gases (eight electrons in the valence shell) the Lewis base, as the donor of a pair of electrons, has in general fulfilled the electron requirements of its own valence shell and consequently possesses pairs of electrons either not being used in bonding, or that can preferentially be transferred for sharing purposes from one attached species to another. The Lewis acid, on the other hand — as an acceptor of a pair of electrons has, in general, yet to satisfy the maximum requirements for electrons its valence shell, and is thus able to provide accommodations for and, consequently, a pair of electrons.

It may be observed that in no wise does the Lewis concept of acid or base alter the classifications made of acids and bases in the Brönsted-Lowry system. All Brönsted-Lowry bases of aqueous function are Lewis bases as well, because in accepting H^+ they conform to the Lewis definition of being electron-pair donors. Likewise, the species classified as acids in the Brönsted-Lowry concept, by virtue of their ability to donate H^+, remain as acids in Lewis nomenclature because their common ingredient, H, can function as an electron-pair acceptor.

The Lewis delineations augment the category of bases with a multitude of chemical substances that in nonaqueous reactions reveal themselves to be electron-pair donors. Likewise, not only does it supplement the category of acids by including all nonaqueous electron-pair acceptors but also extends to a very large number of aqueous reaction species this same conforming designation despite their lack of protonic character.

Returning to the latest equations given, the over-all resemblance of net reaction mechanisms within this framework of coordinate covalent donation or acceptance of electron-pairs falls sharply into focus. Thus, for both the aqueous and the nonaqueous reactions depicted the results read empirically:

and

Pondering these Lewis pictures a bit longer, the conclusion is inevitably reached that there is nothing inherently different in the net structural mechanisms of any of these acid-base reactions from those of the simple neutralizations credited to, and familiarly encountered in, aqueous environments. Hence, all species in all media are effectively neutralized when they have fulfilled their maximum valence shell requirements by coordinating electron-pairs.

Projecting the following simple, empirically-written neutralizations:

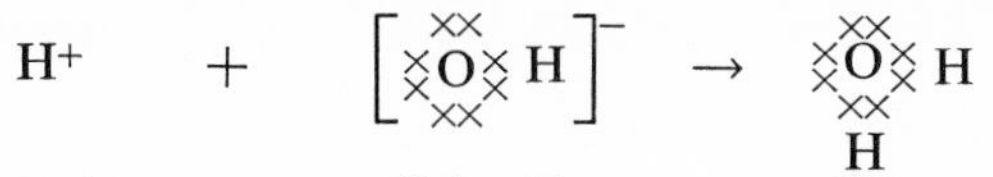

Hydrogen ion	Hydroxyl ion	Water	
Lewis acid	Lewis base	acid-base	
electron-pair acceptor	electron-pair donor	coordination	

and

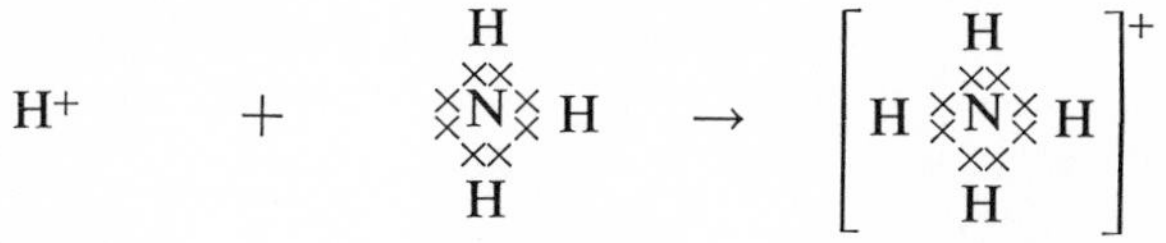

Hydrogen ion	Ammonia	Ammonium ion
Lewis acid	Lewis base	acid-base
electron-pair acceptor	electron-pair donor	coordination

into the familiar acid-base conjugations of ionization, hydrolysis, and amphoterism reveals the essential unity and breadth of application of this theme, and the obvious, purely nominal connotations that are attached to such terms. Thus, the ionizing of the molecule of HCl:

$$HCl + H_2O \rightarrow H_3O^+ + Cl^-$$

represents itself electronically as

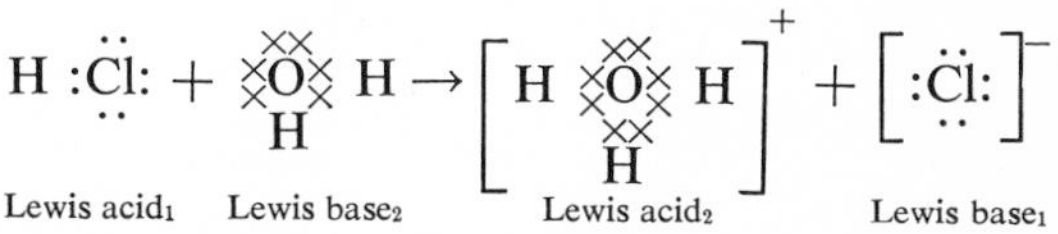

Lewis acid₁   Lewis base₂   Lewis acid₂   Lewis base₁

The hydrolysis of NH_4^+ ion:

$$NH_4^+ + H_2O \rightleftharpoons H_3O^+ + NH_3$$

is represented by

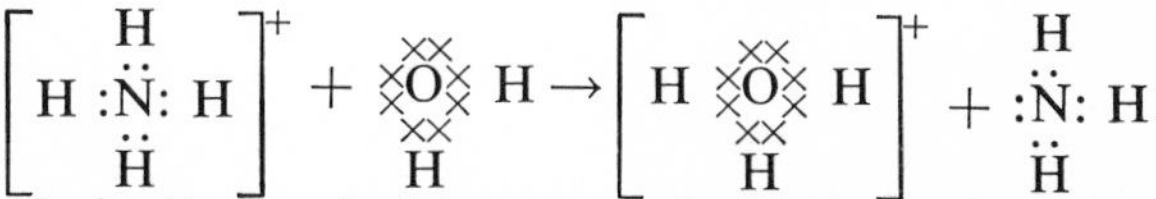

Lewis acid₁   Lewis base₂   Lewis acid₂   Lewis base₁

The way the many amphoteric substances (not merely the hydroxides of almost exclusive classical examples of solubilization) can function in their dual roles of acids and bases reveal, likewise, their identities as donors and acceptors of electron-pairs, whether they be molecular or ionic species. The following equations illustrate the conformity of amphoteric behavior to Lewis definition:

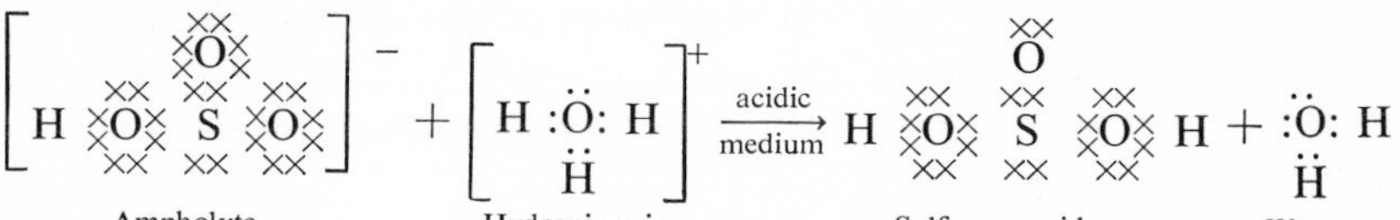

Ampholyte	Hydronium ion	Sulfurous acid	Water
hydrogen sulfite ion	Lewis acid	acid-base	
Lewis base		coordination	

$$\left[H:\ddot{O}:\overset{\displaystyle :\ddot{O}:}{S}:\ddot{O}:\right] + \left[\overset{\times\times}{\underset{\times\times}{\times}\!O\!\times}\, H\right] \xrightarrow[\text{medium}]{\text{basic}} \left[:\ddot{O}:\overset{\displaystyle :\ddot{O}:}{S}:\ddot{O}:\right]^{2-} + \overset{\times\times}{\underset{\times\times}{\times}\!O\!\times}\!\begin{smallmatrix}H\\ \\H\end{smallmatrix}$$

Ampholyte	Hydroxyl ion	Sulfite ion	Water
hydrogen sulfite ion	Lewis *base*		*acid-base*
Lewis *acid*			*coordination*

Transference of H^+ from H_3O^+ to preferential coordination with the HSO_3^- occurs in acidic medium because H_3O^+ is a stronger Lewis acid than is HSO_3^-. Transference of H^+ from HSO_3^- to preferential coordination with OH^- in basic medium occurs because OH^- is a stronger Lewis base than is HSO_3^-.

Perhaps the most profitable applications that are made of the Lewis concepts of acids and bases are in those areas of amphoterism and the formation of complex ions, where transfers or other involvements of protons are either not readily discernible or are expediently omitted when they represent subordinate considerations in the essential chemistry of other species that can be conveniently portrayed in simpler formulations. A few equations serve to illustrate the point:

$$\underset{\text{Lewis }acid}{Al(OH)_3} + \underset{\text{Lewis }base}{OH^-} \rightarrow \underset{\text{Acid-base coordination}}{Al(OH)_4^-}$$

$$\underset{\text{Lewis }acid}{SnS_2} + \underset{\text{Lewis }base}{S^{2-}} \rightarrow \underset{\text{Acid-base coordination}}{SnS_3^{2-}}$$

$$\underset{\text{Lewis }acid}{Cd(CN)_2} + \underset{\text{Lewis }base}{2CN^-} \rightarrow \underset{\text{Acid-base coordination}}{Cd(CN)_4^{2-}}$$

$$\underset{\text{Lewis }acid}{AgCl} + \underset{\text{Lewis }base}{2NH_3} \rightarrow \underset{\text{Acid-base coordination}}{Ag(NH_3)_2^+} + Cl^-$$

$$\underset{\text{Lewis }acid}{Fe^{2+}} + \underset{\text{Lewis }base}{NO} \rightarrow \underset{\text{Acid-base coordination}}{Fe(NO)^{2+}}$$

$$\underset{\text{Lewis }acid}{Ni^{2+}} + \underset{\text{Lewis }base}{6NH_3} \rightarrow \underset{\text{Acid-base coordination}}{Ni(NH_3)_6^{2+}}$$

EXERCISES

Additional factual information is frequently presented in the form of introductory statements to exercises. They should not be overlooked.

1. *Oxyacids* are those that contain the OH group (not the OH^- ion that characterizes bases) which as a unit is bonded covalently to some central atom. Oxyacids may be molecular or anionic. The acidities of aqueous solutions of oxyacids result from the ruptures of O—H bonds in the course of reaction with the solvent water molecules; thus

$$XOH + H_2O \rightarrow H_3O^+ + XO.$$

Although all oxyacids must contain at least one hydroxy group linked to the central atom, the direct attachments to the central atom of lone oxygen atoms and of lone hydrogen atoms are by no means precluded. A few serviceable generalizations may be empiricized from experimental observations that helpfully corroborate the structures of oxyacids; these are:

It is solely the H of the OH group that undergoes ionization in aqueous media. Any lone centrally linked hydrogen atoms make no contributions to aqueous acidity because they are not involved in protolysis; that is, the donation of H^+ ions to the H_2O molecules.

The relative strengths of oxyacids may be approximated in terms of the lone oxygen atoms linked to the central atom. With acidities to be evaluated from the respective equilibrium constants of acid ionization

$$\frac{[H^+][XO^-]}{[XOH]} = K_a$$

The approximate responses of oxyacids are as follows:
— with no oxygen atoms present, $K_a \sim 10^{-8}$ to 10^{-10}; values represent very weak acids
— with just one lone oxygen atom present, $K_a \sim 10^{-2}$ to 10^{-4}: values represent moderately weak acids
— with two or with three lone oxygen atoms present, $K_a \sim 10^1$ and larger (generally very much so): values represent strong acids.

(a) In terms of the foregoing considerations, provide the correct attachments of the bonded atoms in each of the following three oxyacids of phosphorus, for each of which the approximate extent of primary ionization is of the order of $K_a \sim 10^{-3}$: H_3PO_4 (orthophosphoric acid); H_3PO_3 (phosphorous acid); H_3PO_2 (hypophosphorous acid).

(b) In the absence of any information with respect to extents of aqueous acidity and/or K_a values, what alternate structural formulations might have been conceived for the oxyacid, H_3PO_3? In terms of the aqueously ionizable hydrogen, label each of the structural forms descriptively as "monoprotic," or "diprotic," or "triprotic," as is applicable.

(c) State whether the following projected aqueous ionization will occur, supplying as well in each instance an explanation of your answer:

(1) $H_2PO_2^-$ (hypophosphite ion) $+ H_2O \rightarrow H_3O^+ + HPO_2^{2-}$

(2) $H_2PO_3^-$ (phosphite ion) $+ H_2O \rightarrow H_3O^+ + HPO_3^{2-}$

2. Comparisons of oxycompounds of central atoms in the same horizontal sequence of periodic classification and which, additionally, are in their states of maximum oxidation, reveal that as the oxidation number of the central atom increases, the acidity (acid ionization) of the oxycompound also increases. A plausible interpretation of this effect is that as the positive nuclear charge of the central atom grows, the greater becomes its pull upon the electron-pair that it shares with the oxygen atom(s). This stress of shifting the density of electrons toward the central atom is, in turn, transmitted to the oxygen atom(s), which must likewise now exert a stronger pull upon the electron-pairs that are being shared with the hydrogen atom(s). Each hydrogen atom thus affected is now

placed at a greater disadvantage in retaining the atmosphere of electrification conducive to its stability within the acid molecule and, consequently, it will avail itself of readier opportunities to do so by reacting with basic species that make available more easily the requisite electron-pairs.

In conformity with the preceding considerations, arrange the following oxycompounds in their correct order of increasing acidic strength: $(HO)_3PO$, $(HO)ClO_3$, $Al(OH)_3$, $Mg(OH)_2$, $NaOH$, $Si(OH)_4$, $(HO)_2SO_2$.

3. For elements capable of forming a series of electronically similar oxycompounds, in which all oxygen atoms are attached to the central element, the following rules are applicable:

   The higher the state of oxidation of the central element, the greater the acidic strength (the smaller the basic strength) of the compound.

   The greater the number of lone oxygen atoms attached to the central atom, the greater the acidic strength (the smaller the basic strength).

   (a) Arrange in order of decreasing acidic strength the members of the following series: H_3PO_2, H_3PO_4, H_3PO_3.

   (b) Arrange in order of decreasing acidic strength the members of the following series: $HClO_4$, $HClO_2$, $HClO$, $HClO_3$.

   (c) It is given that H_3AsO_4 provides no (zero) lone oxygen atoms in the structure of its oxycompound. Write molecular formulas that indicate correctly the numbers of lone oxygen atoms and of hydroxyl groups for the following series wherein acidities are decreasing in the following order: $HClO_4 > H_2SO_4 > H_2SO_3 > H_3AsO_3$.

   (d) What explanation plausibly interprets rule two of the introductory statement provided with this exercise?

4. The acidity of structurally-similar oxycompounds of central atoms in the same chemical family decreases downwards in their respective periodic groups. Consequently, basic strength must increase in the same direction.

   Arrange, in order of increasing basicity, the following: $Bi(OH)_3$, $As(OH)_3$, $Sb(OH)_3$.

5. For central atoms in the same state of oxidation, the smaller their sizes the greater the density of their charges (charge/unit volume). Based upon this principle, the following rules have been found to apply:

   1. With respect to oxyacids of molecular constitution wherein the respective central atoms are all in the same state of oxidation: The weakening of the H—O bond that favors increased acidic ionization is promoted by increased density of charge (diminished size of central atom). This may be interpreted as the increased pull exerted by the central atom upon the electron-pair of the H—O bond. Conversely, the greater the dispersal of the charge (enlarged size of central atom) the more pronounced the basic properties.

   2. With respect to binary (just two elements) molecular acids of nonoxy constitution: The smaller the size of the central atom (when in the same periodic group), the greater becomes the increment of negative charge upon the central atom, and the greater, concomitantly, becomes the increment of positive charge upon the hydrogen atom in their mutual covalent linkage. This means that the percentage of ionic character of the bond increases

$$\rightarrow \left[\begin{array}{c} Ca^{2+} \\ 0.25 \text{ g-formula } Ca^{2+}/\text{liter} \\ \left(= 0.25 \text{ mole } Ca^{2+}/\text{liter}\right. \\ \left.= 0.25F\ Ca^{2+} = 0.25M\ Ca^{2+}\right) \end{array}\right] + \left[\begin{array}{c} 2Cl^- \\ 0.50 \text{ g-formula } Cl^-/\text{liter} \\ \left(= 0.50 \text{ mole } Cl^-/\text{liter}\right. \\ \left.= 0.50F\ Cl^- = 0.50M\ Cl^-\right) \end{array}\right]$$

MOLALITY

Molality (m) means the number of moles of solute dissolved in 1000 grams of solvent; that is, $m = $ moles of solute in 1000 g of solvent. Concentrations of solution expressive of molality are immune to the temperature-dependent variations to which volumes are subject. Some qualification, however, is necessary with respect to the extent to which the ionization or dissociation of a solute is altered by variation in temperature, which, as a result, may lead to significant differences in the component units derived from the cleavage of the parent solute. Weights of nondissociable and completely ionic species otherwise remain constant in all conventional laboratory work. A 0.10 molal solution of glucose thus remains dependably 0.10 m $C_6H_{12}O_6$ for all occasions. Under ordinary conditions, when dealing with dilute aqueous solutions, concentrations by *molality* and by *molarity* are virtually the same. This is because of the nearly numerical equivalence of 1000 g of H_2O to 1000 ml of H_2O as solvent, and (considering the relatively small quantities of solute generally present) of 1000 ml of solvent H_2O to 1000 ml of the final solution.

Although molality has proved to be readily applicable in the measurement and calculation of freezing points, boiling points, vapor pressures, and osmotic pressures of solutions, as a concentration for experiments it is ordinarily better avoided because of the inconvenience of weighing liquid solvents. The following progression interprets appropriately the concentrations of component ions of solute species in reference to the parent solute in accordance with which ambiguity may be avoided. We will make an aqueous solution of 0.25 m $CaCl_2$:

$$\left[\begin{array}{l} CaCl_2 \text{ (solid)} \\ 0.25 \text{ g-formula } CaCl_2 \\ (= 0.25 \text{ mole } CaCl_2) \end{array}\right]$$

$$\rightarrow \left[\begin{array}{c} CaCl_2 \text{ (dissolved parent conc.)} \\ 0.25 \text{ g-formula } CaCl_2/1000 \text{ g } H_2O \\ (= 0.25m\ CaCl_2) \end{array}\right]$$

$$\rightarrow \left[\begin{array}{c} CaCl_2 \text{ (existent as such)} \\ 0.00 \text{ mole } CaCl_2/1000 \text{ g } H_2O \\ (= 0.00m\ CaCl_2) \end{array}\right]$$

$$\rightarrow \left[\begin{array}{c} Ca^{2+} \\ 0.25 \text{ mole } Ca^{2+}/1000 \text{ g } H_2O \\ (= 0.25m \ Ca^{2+}) \end{array} \right]$$

$$+ \left[\begin{array}{c} 2Cl^- \\ 0.50 \text{ mole } Cl^-/1000 \text{ g } H_2O \\ (= 0.50m Cl^-) \end{array} \right]$$

MOLE-FRACTION

Mole-fraction (x) means the ratio of the number of moles of a particular solute present in a solution to the total number of moles of all components of the solution, including the solvent.

For a solution composed of n_1 moles of solvent and n_2 moles of its single solute, the mole-fraction (x_1) of the solute is calculated as

$$x_1 = \frac{n_2}{n_1 + n_2}.$$

Correspondingly, for the system of n_1 moles of solvent in which a *multiple* number (i) of component solutes have been dissolved in the relative numbers of moles, respectively, of n_2, $n_3 + \cdots + n_i$ the mole-fraction of the component represented by n_2 moles is

$$x_2 = \frac{n_2}{n_1 + n_2 + n_3 + \cdots + n_i}.$$

Mole percentage, as sometimes used, involves merely the multiplication of the mole-fraction by 100.

The sum of all of the mole-fractions of all components of a solution must necessarily be equal to unity; thus, for a three-component system,

$$x_1 = \frac{n_1}{n_1 + n_2 + n_3}; \ x_2 = \frac{n_2}{n_1 + n_2 + n_3}; \ x_3 = \frac{n_3}{n_1 + n_2 + n_3}$$

and

$$x_1 + x_2 + x_3 = \left(\frac{n_1}{n_1 + n_2 + n_3} \right) + \left(\frac{n_2}{n_1 + n_2 + n_3} \right) + \left(\frac{n_3}{n_1 + n_2 + n_3} \right)$$

$$= \frac{n_1 + n_2 + n_3}{n_1 + n_2 + n_3}$$

$$= 1.$$

The utility of *mole-fraction* as a unit for solutions is of particular significance in quantitative calculations dealing with colligative properties of solutions. The avoidance therein, by use of this unit, of the volume variations resulting from changes of temperature (and, sometimes of pressure) is a "must" if the properties of solutions that depend on concentration — boiling point, freezing point, vapor pressure, osmotic pressure — are to be evaluated correctly.

WEIGHT-FRACTION

Weight-fraction means the ratio of the number of parts by weight of a particular solute dissolved in a given sample of solution to the total weight of the sample (solvent plus all component solutes). The considerations applied to mole-fraction apply similarly to weight-fraction (and to weight-percentage), the sole distinction here being the choice of a physical unit, generally *gram*, for the solute rather than the chemical unit of mole.

Caution must always be exercised to avoid the ambiguity caused by use of the term "solute-percent," inasmuch as concentration of solute may likewise be placed upon a basis of volume of solution. This could mean either the weight of solute in a unit volume of total solution (weight-volume fraction; and percent) or volume of solute in a unit volume of total solution (volume-volume fraction; and percent); or even, confusingly enough, either weight or volume of solute to volume of solvent. Such variations are virtually lacking in quantitative chemistry, but they are so frequently used in biology and pharmacology that the required term ought to be spelled out fully.

NORMALITY

Normality (N): means the number of gram-equivalent weights of solute contained in one liter of a final solution; that is, $N =$ gram-equivalents of solute to 1000 ml of solution. The concept of normality of a solution will prove especially useful in quantitative calculations dealing with volumetric stoichiometry. We, therefore, defer elaborations and applications of this unit to that stage of our expositions. We cite here, however, the especial service that normality renders to the precise understanding of the nature of reactions in solution by precisely identifying the chemical species that are actually responsible. Excluded, thus, are the "spectator" species that are not involved, but which, normally, are also introduced by the parent solute as a whole.

Also to be cited is the relationship of the normality to the molarity of a solution. The normality of any solution must always be a whole-number multiple of its molarity. This follows directly from the fact that the gram-equivalent weight of the solute — the numbers of which constitute and determine the numerical value of the normality — is itself a quotient of the molar (gram-formula) weight, divided by some usually small whole number

$(1, 2, 3, 4, 5 \ldots)$. Hence, a specific $1.0M$ solution may be $1.0N$, or $2.0N$, or $3.0N$, or larger, as may be observed in the chapter dealing with volumetric stoichiometry, where gram-equivalent weights are computed not for acids, bases, and salts, but for oxidants and reductants.

In redox, as already described, gram-equivalent weights of reactant species are fractional parts of moles, determined not by valences (normally 1, 2, or 3) but rather by electrons lost (in oxidation) and gained (in reduction).

When the latter species is complex, these values prove frequently to be numerically larger than the valences of the respective species. For example, the $Cr_2O_7^{2-}$ ion gains six gram-electrons per mole (six electrons per ion) in being reduced to the Cr^{3+} ion; and the MnO_4^- ion gains five gram-electrons per mole (five electrons per ion) in being reduced to the Mn^{2+} ion. On the other hand, the transfer of electrons may well be smaller than the valence of the species; for example, conversion of the Fe^{2+} ion to the Fe^{3+} ion, or vice versa, involves the transfer of only one gram-electron per mole (one electron per ion).

In any event, the normality of the solution of a reactant cannot be reliably foretold from its known molarity without a precise stipulation of the chemical identity of the product(s) resulting from the reaction. It is only from such information concerning the alternative products of reactions that are possible under differing experimental conditions that we may correctly establish the multiples of whole numbers necessary to transpose molarities of solutions to normalities; and likewise, gram-formula (or molar) weights to gram-equivalent weights.

Parenthetically, the symbol N, in the context of a concentration term, must not be confused with the identical symbol representative also of the Avogadro number.

DENSITY AND SPECIFIC GRAVITY

Density and specific gravity are not normally regarded as units of concentration of a solution. The exposition here of their individual significance and mutual interrelationships are merely concessions to the occasional needs of students in experimental work to translate the data provided on the labels of reagent bottles to the concentration units already presented (molarity, formality, and normality in particular).

Density and specific gravity are by no means the same, although they are frequently confused. Perhaps the source of this confusion is the fact that numerically the specific gravity of any solution prepared by dissolving a solute in a liquid solvent is identical with the mass of an equal volume of pure H_2O at $4°C$ (the common reference standard). This follows, necessarily, from the definition of the specific gravity of a solution as the ratio of its density (mass per unit volume) to the density of the standard with which it is being compared. As the standard of H_2O at 4°C has a mass of 1.000 g/ml, the numerical identity of the two terms becomes inescapable; thus

$$\text{specific gravity of solution} = \frac{\text{density of solution in grams/ml}}{1.000 \text{ g } H_2O/\text{ml}}.$$

To be noted, however, is the dimensionless character of specific gravity. With the density units of grams/ml cancelling out in the ratio, specific gravity becomes a pure number. Any factor that changes the density of the standard from a numerical value of unity will operate to create a numerical divergence between specific gravity and density. In laboratory practice, the standard of reference may be H_2O at some temperature other than 4°C; and, indeed, the temperature frequently noted is 25°C.

Clearly, the density of a solution of H_2SO_4 whose label bears the notation

$$\text{sp.gr.} = 1.840_{25°}^{25°}$$

would not, at 25°C, be equal to 1.840 g/ml, inasmuch as the density of pure H_2O at 25°C is not 1.000 g/ml but, rather, 0.9971 g/ml. Were the label to read, however,

$$\text{sp.gr.} = 1.840_{4°}^{25°}$$

it could be accepted that the density of the H_2SO_4 solution at 25°C (as indicated by the *super*script is 1.840 g/ml when compared to that of H_2O at 4°C (as indicated by the *sub*script). The deviations may be slight enough to be negligible in comparsions of specific gravity and density for aqueous solutions, but emphatically not when other solvents are employed: or when, as in some instances, the metric system of comparisons is not being used. In the latter case, the numerical identity of specific gravity with density vanishes even when reference is made to H_2O at 4°C. This necessarily follows from utilizing the conversion factors requisite to changing the metric units of grams and milliliters to the nonmetric units of ounces, pounds, cubic inches, cubic feet, etc.

Unlike density, specific gravity remains unalterably constant regardless of the system of measurement used. Thus, in the metric system both the specific gravity and the density of metallic mercury, Hg^0, are numerically 13.6. In the avoirdupois system, however, the metal's density must be described as 7.85 (avoir.) ounces to the cubic inch while its specific gravity still remains constant at 13.6.

Appropriate computations employing specific gravity and density of solutions are offered in the chapter dealing with volumetric stoichiometry.

CHEMICAL EQUIVALENCE — UNITS OF METATHESIS AND REDOX

Reaction always occurs equivalent for equivalent. The computative convenience of this concept in establishing the relationships of weight and volume of chemically interacting substances, as well as of the products of their reaction, has already been discussed and cannot be too strongly stressed. It nearly always spares the student of chemistry the need to work out a balanced or a partially balanced equation for subsequent purposes of "proportionating" the molar weights of the materials involved in interaction. The specific common basis of this concept of chemical equivalency, which universally integrates the factors of quantity in all chemical reactions — regardless of their nature, whether metathetical (nonredox) or redox — is, once again, the Avogadro number, 6.02×10^{23}. On this basis we validly interpret the requirements of weight and/or volume for the various classes of chemical reactants in aqueous solutions.

The following definitions and interpretations are in order:

1. *Gram-equivalent weight of an acid*: That weight of an acid that yields to a base one mole of H^+ ions; (that is, 6.02×10^{23} ions of hydrogen).

2. *Gram-equivalent weight of a base*: That weight of a base that (if an ionic hydroxide) will deliver to an acid one mole of OH^- ions (that is, 6.02×10^{23} hydroxyl ions); or, if the base be an uncharged molecular covalent substance, that weight thereof that accepts from an acid one mole of H^+ ions.

The intent of equivalents here is unambiguous. It takes 6.02×10^{23} ions of H^+ to react with 6.02×10^{23} ions of OH^- to produce, in their mutual mole-for-mole neutralizations, 6.02×10^{23} molecules of H_2O. Hence, the weights of acids and bases that react in neutralizations are those that supply these requisite molar quantities of ions. It is clear that diprotic and triprotic acids or bases are capable of providing more than one equivalent weight, depending upon the demands of reaction. Thus, although the equivalent weight of HCl can be only its molar weight, the equivalent weight of H_3AsO_4 (arsenic acid) may be *variably*,

(a) A full molar weight if the reaction is restricted to

$$H_3AsO_4 + OH^- \rightarrow H_2AsO_4^- + H_2O$$

(b) One-half the molar weight if reaction is restricted to

$$H_3AsO_4 + 2OH^- \rightarrow HAsO_4^{2-} + 2H_2O$$

(c) One-third the molar weight if reaction involves a full and complete neutralization of all available hydrogen of the acid; that is,

$$H_3AsO_4 + 3OH^- \rightarrow AsO_4^{3-} + 3H_2O$$

Similar appraisals can be made for bases. What must be stressed here is that, unlike the rigid constancy of molar weight, the gram-equivalent weight may prove to be a variable factor, inasmuch as it is not the mere potentiality of a species to make available H^+ or OH^- ions. Rather, it is its actual delivery of H^+ or OH^- in terms of the demands made upon it by the "opposite" reactant. Just how much H^+ aon has to be delivered to a base by a polyprotic acid depends upon the quantity of base present to be neutralized. Likewise, the amount of OH^- ion that has to be delivered to an acid by an alkaline-earth hydroxide (e.g., $Ca(OH)_2$) or the amount of H^+ ion that a covalent molecular base can accept depends upon the quantity of acid introduced.

3. *Gram-equivalent weight of a salt*: As salts are merely aggregates of cations other than H^+ and anions other than OH^-, we may for our purposes figuratively transpose the cation and the anion to terms of their "replaceabilities" by H^+ and OH^- ions, respectively.

For example, the cation Mg^{2+} in the salt $MgSO_4$, being replaceable by two H^+ ions, and the anion SO_4^{2-}, being likewise replaceable by two OH^- ions, the requirement that equivalent weight conform to the delivery of 6.02×10^{23} particles of each will be fulfilled by one-half of the molar weight of the salt.

The equivalent of any simple salt must then be, inescapably, the molar weight divided by the total valence either of the anion or of the cation. In either case, the total valence of the one must be numerically identical with the total valence of the other — in a simple salt that is, wherein there is only one kind of anion and one kind of cation; and, hence, the quantity of the salt representing the equivalent weight must be the same for both. In further illustration, the equivalent weight of K_2SO_4 would be one-half of the molar weight of the salt; that of Na_3PO_4 would be one-third of the molar weight of the salt.

When salts are the double or complex type — that is, having more than one kind of cation and/or of anion — no difficulties in computation of equivalent weights need be experienced. All that is necessary is to specify the particular ionic constituent that is undergoing reaction or being prepared for reaction, and to follow the same procedure; namely, to compute the gram-equivalent weight of such a salt by dividing the molar weight by the total valence of the ionic constituent of interest in the reaction. Thus, the gram-equivalent weight of $NaAl(SO_4)_2 \cdot 12H_2O$ will be the full molar weight of the entire salt (including the $12H_2O$) when the Na^+ ion is stipulated (total valence, 1); one-third of the molar weight of the salt when the Al^{3+} ion is stipulated (total valence 3); and one-fourth of the molar weight of the salt when the SO_4^{2-} ion is stipulated (total valence, 4).

4. *Gram-equivalent weight of an oxidizing agent*: That weight of any substance which, in chemical reaction, acquires from a reducing agent the

Avogadro number of electrons; that is, 6.02×10^{23} electrons — the gram–electron.

5. *Gram-equivalent weight of a reducing agent*: That weight of any substance which, in chemical reaction, gives up to an oxidizing agent the Avogadro number of electrons; 6.02×10^{23} electrons — the *gram–electron*.

The implications of the extent of electron-transfers between species that determine the precise nature of oxidation-reduction products and the variability of reactant equivalents receive extended treatment in later chapters on their applications to the stoichiometry of solutions.

CHAPTER 4

CONDUCTANCE OF IONS

We, first, define a number of the electrical terms and units to be encountered throughout this chapter.

ELECTRICAL TERMS AND UNITS

Coulomb — quantity of electricity representing the charge required to reduce 0.001118 g of dissolved Ag^+ ion to metallic Ag^0, with the deposition of this weight of the latter upon a cathode.

Faraday — collective electrical charge equal to that yielded by the Avogadro number of electrons; that is, the gram-electron, or 6.023×10^{23} electrons. As the charge upon an individual electron is experimentally established as 1.60×10^{-19} coulombs, the total charge represented by the faraday is

1.60×10^{-19} coulomb/electron $\times 6.0235 \times 10^{23}$ electrons/faraday

$= 96,500$ coulombs/faraday (as rounded off)

Ampere — amount of current flow equal to the transport of one coulomb of charge a second past a given point in a conductor.

Ohm — amount of electrical resistance equal to the resistance offered to the transport of current at $0°C$ by a column of pure metallic mercury 106.30 cm long and 1 sq cm in cross section.

Resistance varies directly as the length of the conductor and inversely as its cross-section. Tripling the length of the conductor, for example, triples its resistance; tripling its cross-section, on the other hand, reduces resistance to one-third of what it was originally. The resistance of a conductor also depends upon the chemical nature of the substance itself; hence, we refer to *specific resistance* to the transport of electricity as evaluated for any given conductor. This is defined as the resistance, in ohms, of a cube of the conductor, one-centimeter on edge, to electrical conductance from any one of its faces to an opposite face. Specific conductance is frequently designated by the Greek letter, rho, ρ.

Volt — amount of electrical pressure (electromotive force, or difference of potential) that must be applied to a conductor in order to convey a current flow of one ampere through a resistance of one ohm. The definitive mathematical relationship for the flow of direct current is expressed by Ohm's law, namely

$$\text{volts} = \text{amperes} \times \text{ohms}.$$

With any two of these terms known, the third is immediately determined.

Joule — amount of electrical work, representing the energy expended during the flow of a direct current of one ampere through a resistance of one ohm for a period of one second. A useful definitive relationship expressing the joule in its composite units of *volt-coulombs* is as follows:

$$\text{joules (volt–coulombs)} = \text{electromotive force (volts)} \times \text{quantity}$$
$$\text{of electricity (coulombs)}.$$

Watt — The unit of electrical power, of especial interest to the industrial chemist concerned with the quantity of electrical energy consumed and its related costs. It is the product of one volt times one ampere; hence

electrical power (P) in watts
$$= \text{electromotive force } (E) \text{ in volts} \times \text{current } (i) \text{ in amperes}.$$

One horsepower is equal to 746 w. Watts, when multiplied by time in hours yields the familiar *kilowatt hours* representing the usual base for calculating costs in the industrial or domestic consumption of electricity.

Reciprocal ohm (mho) — quantity of electrical conductance which for any specific conductor — metal or electrolyte — is equal to the quotient of 1/specific resistance, in ohms.

As specific resistance defines the resistance of a cube of the conductor, one centimeter on edge, to the flow of current between two opposite faces, clearly the reciprocal ohm (or mho) represents *specific conductance*.

Specific conductance in reciprocal ohms is frequently designated by the Greek letter, kappa, κ. Although this term is valuable as an expression of the total number of electrical charges present in a cubic centimeter of conductor, specific conductance is not particularly informative of the interactions among these electrical charges. If interionic electrical forces, mobility of ions, and particularly the effects of dilution upon the capacity of different solutions to conduct an electrical current, are to be plausibly evaluated, clearly it will be necessary to maintain equal numbers of charges for purposes of valid comparison.

As this condition is not obtainable with values of specific conductance, we therefore apply the concept of *equivalent conductance*, conventionally designated by the Greek capital letter, (lambda Λ). This represents the conductance of a constant number of electrical charges, specifically that of one gram-equivalent weight of an electrolyte solute, in a solution that is placed between two parallel flat electrodes, spaced precisely one centimeter apart, and which are of sufficient size to contain the entire solution. The constancy of the number of electrically charged particles obtained from equal numbers of gram-equivalent weights at "infinite" dilution (100% dissociation of the solute is to be presumed) should be fully appreciated.

In illustration, a mole of $K_3Fe(CN)_6$, in yielding potassium and ferricyanide ions, contains three faradays each of positive and negative charge; a mole of $CuSO_4$, in yielding copper and sulfate ions, contains two faradays each of positive and negative charge; and a mole of NaCl, in yielding sodium and chloride ions, contains one faraday each of positive and negative charge. Hence, the weights of these solutes that provide a numerical identity and constancy of charge for comparative purposes are, respectively, one-third of the gram-formula weight of $K_3Fe(CN)_6$, one-half the gram-formula weight of $CuSO_4$, and one gram-formula weight of NaCl. Comparative evaluations of interionic forces of attraction, ionic, mobility, and effects of dilution are thus made in terms of equal numbers of equivalent weights

In practice one does not actually measure equivalent conductance directly as such; rather it is calculated from the specific conductance of one cubic centimeter of solution of known normality (and as already defined by kappa, κ), multiplied by the volume in cc (v) required to contain the full gram-equivalent weight of the solute.

Thus, for an electrolyte of known normality,

equivalent conductance (Λ) = specific conductance (κ) $\times$ volume (v)

or, as one cc of solution of any given normality (N) must be multiplied by $1000/N$ in order to make sufficient solution to contain one full gram-equivalent weight of solute, we may also express the required relationship of equivalent conductance to specific conductance by:

$$\Lambda = \kappa \times \frac{1000}{N}.$$

A simple mathematical illustration crystallizes all necessary facets of interpretation. If, for example, the specific resistance of 0.10 N $HC_2H_3O_2$ at 18°C is determined as 2173.9 ohms,

$$\Lambda\left(\begin{array}{c}\text{equivalent}\\\text{conductance}\end{array}\right) = \frac{1}{2173.9}\left(\begin{array}{c}\text{specific}\\\text{conductance}\end{array}\right) \times \frac{1000}{0.10}$$

and

$$\Lambda 0.10 N \ HC_2H_3O_2 = 4.60 \text{ reciprocal ohms (mhos).}$$

If the quotient $1000/N$ (in this instance, $1000/0.10$) be not immediately comprehended as a conversion factor necessary to provide one full gram-equivalent weight of the acid, observe the valid proportion:

1 cc of 0.10 N $HC_2H_3O_2$ supplies 0.00010 g-equiv of solute
1/0.00010 cc supplies 1 g-equiv of solute
1/0.00010 = 1000/0.10 (= 1000/N) cc supplies 1 g-equiv of solute.

It is apparent that the conductance of a solution may also be expressed as molar conductance and equated, when desired, in a manner similar to that of equivalent conductance; thus

molar conductance (μ) = specific conductance (κ) × volume (cc)
required to contain a full mole of the electrolyte

whence

$$\mu = \kappa \times 1000/M,$$

bearing in mind the numerical inequalities of the charges existing in molar solutions of electrolytes of different valences; for example, A^+B^-, as compared with $C^{2+}D^{2-}$, $C^{2+}(B^-)_2$, or $(A^+)_2D^{2-}$.

EQUIVALENT CONDUCTANCE AND CONCENTRATION OF IONS

With conventional units now defined, let us investigate the qualitative relationships experimentally observed between conductance and the ionic concentrations of solutions. For any specific temperature the conductance of a solution depends, at least theoretically, upon the total number of ions, their valence (indicates the faraday units of positive and negative charges they carry), and their mobility or speed of movement. Clearly, then, the specific conductance of any strong electrolyte (presumed 100% dissociated) varies in direct proportion to its concentration — making due allowance, necessarily, for the changes in ionic speeds that occur as variability in dilution alters the extent of interionic attraction and the hydration of ions. As the solution of the strong electrolyte is progressively diluted, the specific conductance progressively diminishes and approaches zero because the practical effect of infinite dilution is a "zero concentration" of the solute. Not so, however, with the equivalent conductance of the strong electrolyte; because the definition as made for equivalent conductance fixes it unalterably as a numerical constancy of charge — the gram-equivalent.

Purely from the point of view of constancy of charge, the equivalent conductance of the strong electrolyte ought to be completely independent of concentration. In practice, however, the equivalent conductance of a strong electrolyte is observed to increase with progressive dilution up to the maximum limiting value defined by infinite dilution, Λ_o (equivalent conductance at infinite dilution). This is interpretable, however, not as a change in the numbers of ions or charges — because strong electrolytes are to be regarded as fully ionized at all concentrations — but rather as a progressive release from the electrical inhibitions upon their freedom of migration. Quite logically, as dilution increases the distances between the ions, their mutual attractions must diminish. Experimentally, then, a strong electrolyte validly conforms to a reasonably approximate constancy of equivalent conductance in its entire range of dilute solution. Even in highly concentrated solutions its equivalent conductance is relatively very high as compared to that of the weak electrolyte.

The equivalent conductance of a weak electrolyte likewise increases with progressive dilution, gradually approaching the limiting value of Λ_o, but its ascent in this direction is so very gradual that the slow increases of equivalent conductance with increasing dilution cannot be reconciled with the mere reduction of interionic attractions alone. Rather, this must be attributed predominantly to an actual numerical increase of ions, that is, to greater dissociation or cleavage of the parent electrolyte. Measurements of equivalent conductances thus yield firm interpretative grounds for the contrasts existing between weak and strong electrolytes. The latter are presumed to be present completely in ionic form, even though ion-clustering may significantly retard ion-mobility and, consequently, impair the full experimental manifestations of their presence. The figure illustrates the considerations just developed:

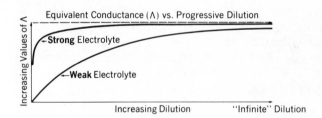

Before proceeding with further considerations of conductance of the electrolytes and mobility of the ions, a few words of explanation are in order with respect to the mechanics of achieving "infinite dilution" experimentally; literally, it is unattainable. What is actually done is to add sufficient water until the progressively measured equivalent conductance, Λ, reaches a value that no longer changes significantly. For strong electrolytes, comprising nearly all the inorganic salts and familiar ionic acids and alkali bases, this near-constancy of equivalent conductance is achieved in approximately 10,000 to 15,000 liters. This permits a practical graphic plotting of values for progressive conductance versus volumes in progressive dilution and the extrapolation from the curve thus obtained of the desired value of Λ_0 — the limiting value of equivalent conductance.

It must be observed, however, that although this device for ascertaining Λ_0 works satisfactorily for highly ionic, strong electrolytes, its application to weak electrolytes yields most inaccurate data inasmuch as weak electrolytes do not permit the linear extrapolations of Λ_0 as do the strong electrolytes. Evan at large dilutions the steady cleavage of un-ionized species into ions may continue significantly. The mobility of all species of a weak electrolyte cannot be presumed to be unimpaired by interionic forces of attraction and repulsion. The interactions of solute and solvent which are responsible for the ionizing cleavage in molecular species, as well as the concomitant, highly variable, hydration (solvation) effects and electrical "drags" upon the ions so formed, present perplexing difficulties in equating the volumes of relatively free solvent. It is the consideration of free solvent that is implicit in any graphic extrapolations of Λ_0 from progressive additions of solvent.

As a reasonably accurate evaluation of Λ_0 is of transcendant importance in calculating the extent of ionization of a weak electrolyte, a different course to this objective must be sought. It is provided, happily, in an extension of the work of F. Kohlrausch (Germany, circa 1875).

KOHLRAUSCH'S LAW OF INDEPENDENT ION MIGRATION

In Kohlrausch's studies of conductance, it is concluded that at infinite dilution the cations and anions of a strong electrolyte move completely independently of one another — in which event its total equivalent conductance must be the sum of the separate equivalent conductances at infinite dilution of its ions. We may thus express Λ_0 in terms of its conductance of constituent ions at the limiting value, employing λ_0 (lower case Greek letter, lambda) to designate separate ionic conductances:

$$\Lambda_0 = \lambda_{0\ \text{cation}} + \lambda_{0\ \text{anion}}.$$

We interpret, illustratively, the implications of this expression and, by subsequent application of pertinent experimental values, its utility in evaluations of Λ_0 of weak electrolytes:

$$\Lambda_0 \text{NaCl} = \lambda_0 \text{Na}^+ + \lambda_0 \text{Cl}^-$$

and

$$\Lambda_0 \text{KCl} = \lambda_0 \text{K}^+ + \lambda_0 \text{Cl}^-$$

Algebraically, we equate

$$\Lambda_0 \text{NaCl} - \Lambda_0 \text{KCl} = \lambda_0 \text{Na}^+ - \lambda_0 \text{K}^+ + \lambda_0 \text{Cl}^- - \lambda_0 \text{Cl}^-,$$

with Λ_0 Cl$^-$ *cancelling out.* In conformity with independent behavior of the ions and their unimpaired contributions to Λ_0 of the electrolyte solute,

we infer that the difference between NaCl and KCl, in equivalent conductances at infinite dilution, is simply the difference between the limiting equivalent conductances of the positive ions, Na^+ and K^+. In similar fashion the difference in Λ_0 between KBr and KI would be the difference of the values of Λ_0 of the respective negative ions Br^- and I^-.
Thus

$$\Lambda_0 \, KBr = \lambda_0 K^+ + \lambda_0 Br^-$$

$$\Lambda_0 \, KI = \lambda_0 K^+ + \lambda_0 I^-$$

whence, algebraically

$$\Lambda_0 KBr - \Lambda_0 KI = \lambda_0 K^+ - \lambda_0 K^+ + \lambda_0 Br^- - \lambda_0 I^-,$$

with $\Lambda_0 \, K^+$ cancelling out.

DEGREES OF IONIZATION OF WEAK ELECTROLYTES — APPLICATIONS OF Λ_0

It must be stressed that the relationships of Kohlrausch's law, as just provided, are valid for completely ionic solutes only at extremely high dilutions. At the concentrations that are ordinarily encountered in actual experimental work the interionic forces of attraction and repulsion are too strong to permit any acceptance of the idea that ions migrate independently of others. At infinite dilution, on the other hand, it is to be presumed that the requisite conditions of complete freedom of ions are attained, not only for strong but also for weak electrolytes. That is, all molecular and/or coordinately or electrostatically bonded ionic species have been completely ionized or dissociated into simple ions, none of which is influenced in its speed or mobility of movement by the presence of any other ion. Applying Kohlrausch's law, then, to weak electrolytes in these circumstances we obtain required valid evaluations of Λ_0.

To determine, consequently, the limiting equivalent conductance of acetic acid, $HC_2H_3O_2$, we conclude, in consonance with what has preceded, that its derivation will respond algebraically to any or all of a number of related expressions for strong electrolytes. Any mathematical deviations among them would be attributable to the experimental sensitivities of the assigned values of Λ_0. Thus, one possible combination that clearly equates to the desired expression:

$$\Lambda_0 HC_2H_3O_2 = \lambda_0 H^+ + \lambda_0 C_2H_3O_2^-$$

is obtained by mathematically introducing the required H^+ in HCl, the required $C_2H_3O_2^-$ in $NaC_2H_3O_2$, and removing the concomitantly introduced but unrequired Na^+ and Cl^- in their additive composite, NaCl. This yields an expression that permits ready substitution of numerical values as experimentally ascertained for the respective strong electrolytes at 25°C:

$$\Lambda_0 HC_2H_3O_2 = \Lambda_0 HCl + \Lambda_0 NaC_2H_3O_2 - \Lambda_0 NaCl$$
$$= 426.2 + 91.0 - 125.6$$
$$= 390.7$$

The application of this value of Λ_0 in ascertaining the degree of ionization of any given solution of $HC_2H_3O_2$ at a similar temperature, and for some finite equivalent conductance, is comprehended from the following relationship:

$$\alpha = \frac{\Lambda}{\Lambda_0}$$

$$= \frac{\text{equivalent conductance at } \textit{finite} \text{ dilution and incomplete ionization}}{\text{equivalent conductance at } \textit{infinite} \text{ dilution and complete ionization}}$$

where α denotes degree of ionization of the weak electrolyte and

$$\frac{\Lambda}{\Lambda_0} \times 100 = \% \text{ ionization.}$$

Knowing any two of these terms permits assigning a number to the third.

The application of Kohlrausch's law of independent ion migration to weak electrolytes, which has just been made, can also serve with usefulness in determining the apparent extent of ionization of strong electrolytes. As revealed by x-ray studies of the solid-state structures of salts, all strong electrolytes must be regarded as 100% ionized. The intense electrostatic attraction, however, among the very large numbers of positive and negative ions that are normally present even in rather dilute solutions of strong electrolytes sharply diminishes the speed and effective manifestation of their presence. This is the Debye-Hückel theory of reconciling the active or effective concentrations (activities) of ions with their *actual numbers*.

With any solution of given concentration of a strong electrolyte the total numbers of ions must clearly remain constant as the solution is progressively diluted, or is progressively concentrated by removal of solvent. The speeds of the ions, however, are increased as dilution takes them farther apart from the retarding effects of attraction; and conversely, their speeds markedly decrease as removal of solvent brings them closer together. The expectation that apparent degrees of ionization would be very much the same for equivalent concentrations of strong electrolytes in identical valence categories (i.e., mono-monovalent, A^+B^-, di-divalent, $C^{2+}D^{2-}$; monodivalent, $(A^+)_2D^{2-}$, etc.) has been borne out experimentally, with smaller apparent values of α noted (and logically reconciled) with ionogens of higher valence for which attractions are greater and, consequently, activities are smaller. This in no wise contradicts the interpretations made of the large differences that frequently are observed in the values of Λ_0 for strong electrolytes despite their occurrence in types in which the valence is identical but, wherein, because of chemical dissimilarities, the densities of charge and the ionic radii are significantly different.

Thus, compare the discrepancies, at 25°C, of LiCl for which $\Lambda_0 = 1150$ units, with HCl for which $\Lambda_0 = 42.62$ units; or $NaC_2H_3O_2$ for which $\Lambda_0 = 91.0$ units with NaOH, for which $\Lambda_0 = 247.8$ units.
On the other hand, compare the periodic-group-related KCl, KBr, and KI with their quite consistent experimental values for Λ_0 of 149.9, 151.9, and 150.4, respectively.

Nonetheless, apparent degrees of ionization (α) of types of strong electrolytes of identical valence and equivalent concentrations are fairly similar even for unrelated ionogens, inasmuch as in the applied expression

$$\alpha = \frac{\Lambda}{\Lambda_0},$$

the value of Λ likewise undergoes proportionate variation.

Tables 4:1 to 4:4 show convenient appraisals of conductances of a number of familiar electrolytes.

TABLE 4:1

Equivalent Conductances at Finite Dilution (Λ, 18°C)

Total Weight (= 1 equivalent)	Electrolyte								
	HNO_3	HCl	H_2SO_4	KOH	KCl	NaCl	$CaCl_2$	$HC_2H_3O_2$	NH_3 aq
$1.0\ N \times 10^3$ cc	310.0	301.0	198.0	184.0	98.2	74.3	67.6	1.3	0.9
$0.10\ N \times 10^4$ cc	346.4	351.4	225.0	213.0	112.0	92.0	88.2	4.6	3.3
$0.010\ N \times 10^5$ cc	365.0	369.3	308.0	228.0	122.4	101.9	103.4	14.3	9.6
$0.0010\ N \times 10^6$ cc	372.9	375.9	361.0	234.0	127.3	106.4	112.0	41.0	28.0

TABLE 4:2

Specific Conductance (κ) versus Equivalent Conductance (Λ)

Total Weight (= 1 equivalent)	KCl(18°C)		KCl(25°C)	
	κ	Λ	κ	Λ
$1.0\ N \times 10^3$ cc	0.0982	98.2	0.1113	111.3
$0.10\ N \times 10^4$ cc	0.01120	112.0	0.01290	129.0
$0.010\ N \times 10^5$ cc	0.001224	122.4	0.001413	141.3
$0.0010\ N \times 10^6$ cc	0.0001273	127.3	0.0001469	146.9
$0.00010\ N \times 10^7$ cc	0.00001290	129.0	0.00001489	148.9

TABLE 4:3

Equivalent Conductances of Strong Electrolytes at Infinite Dilution (Λ_0 18°C)

Electrolyte	Λ_0	Electrolyte	Λ_0
HCl	426.2	$CaCl_2$	135.8
NaOH	247.8	$AgNO_3$	133.4
KBr	151.9	$MgCl_2$	129.4
KI	150.4	KIO_3	127.9
KCl	149.9	NaI	126.9
NH_4Cl	149.7	NaCl	126.5
KNO_3	145.0	LiCl	115.0
$BaCl_2$	140.0	$NaC_2H_3O_2$	91.0

TABLE 4:4

Equivalent Conductances of Ions at Infinite Dilution (Λ_0 25°C)

Cation	λ_0	Anion	λ_0
H^+	349.8	OH^-	197.6
K^+	73.5	$Fe(CN)_6^{4-}$	110.5
NH_4^+	73.4	$Fe(CN)_6^{3-}$	101.0
Pb^{2+}	73.0	SO_4^{2-}	79.8
Ba^{2+}	63.6	Br^-	78.4
Ag^+	61.9	I^-	76.8
Ca^{2+}	59.5	Cl^-	76.3
Sr^{2+}	59.4	$C_2O_4^{2-}$	74.2
Cu^{2+}	54.0	NO_3^-	71.4
Fe^{2+}	54.0	CO_3^{2-}	69.3
Mg^{2+}	53.1	ClO_3^-	64.6
Zn^{2+}	52.8	F^-	55.0
Na^+	50.1	HCO_3^-	44.5
Li^+	38.7	$C_2H_3O_2^-$	40.9

The information afforded by experimental measurements of conductances of solutions provides one of the computative methods for evaluating closely the solubility product constant of a difficultly soluble substance. In a saturated solution of such a solute, the concentrations of the pertinent ions are so very low (important they may be) that we may validly accept the equivalent conductance of the saturated solution to be extremely close to the equivalent conductance at infinite dilution. At this point the presumed "zero" concentration of the solute as a unit is taken as the sum of the limiting or zero concentrations of its ions. Under these conditions, the mobilities of the individual ions are unimpaired by attractional forces, and hence all ions of the difficultly soluble solute are to be regarded as completely independent species, both physically and chemically. Therefore, we may equate for the equivalent conductance (Λ) of a difficultly soluble substance

$$\Lambda_{solute} \approx \Lambda_{0\ solute} \ (= \lambda_{0\ cation} + \lambda_{0\ anion})$$

where Λ_0 represents the limiting conductances of the respective ions of the solute; that is, their ionic mobilities at presumed infinite dilution ("zero" concentration).

The arithmetical means for computing the dissolved concentration of the difficultly soluble substance, whence we derive K_{sp} has yet to be resolved; but the familiar relationship of concentration (normality, N) to Λ_{solute} and to κ_{solute} (the specific conductance) is available:

$$\Lambda_{solute} = \kappa_{solute} \times \frac{1000}{N_{solute}}$$

Clearly, with concentration of the solute so very small, we cannot neglect the contribution of the solvent itself to the over-all specific conductance of the saturated solution — a contribution that, relative to the solute, may be very large indeed. Hence, if we are correctly to evaluate specific conductance (κ) for the highly insoluble solute we must arithmetically equate

$$\kappa_{solute} = \kappa_{saturated\ solution} - \kappa_{pure\ solvent}$$

The specific conductances of the saturated solution of the solute and of the pure solvent used in preparing the solution must be ascertained by experiment. The limiting ionic mobilities that together comprise Λ_{solute}, and which are derived from the many experimental determinations made with highly soluble substances, are conveniently available in reference tabulations. Interpretative calculations involving these principles in the determination of K_{sp} values are included in the mathematical applications now offered of the principles so far delineated.

Sample Calculations

■ PROBLEM 1: To derive the equivalent conductance of an electrolyte from extraneous sources of supply of its ions.

At $t°C$, a $0.0020F$ solution of KNO_3 exhibits an equivalent conductance (Λ) of 140.5 reciprocal ohms (mhos); that of a $0.0020F$ solution of KCl is 145.8 reciprocal ohms; and that of $0.0020F$ solution of HNO_3 is 412.9 reciprocal ohms. Calculate from this information the equivalent conductance of a $0.0020F$ HCl solution.

Solution. Because, at the concentrations provided, all of the given substances may be regarded as fully dissociated into their respective ions, it follows that the total equivalent conductance of each of the different solutions is validly represented by the sum of the independent equivalent conductances of the ions (λ) therein. Hence, for the given solutions:

(a) $[\Lambda 0.0020F\ KNO_3] = \lambda 0.0020N\ K^+ + \lambda 0.0020N\ NO_3^- = 140.5$ mhos
(b) $\quad [\Lambda 0.0020F\ KCl] = \lambda 0.0020N\ K^+ + \lambda 0.0020N\ Cl^- = 145.8$ mhos
(c) $[\Lambda 0.0020F\ HNO_3] = \lambda 0.0020N\ H^+ + \lambda 0.0020N\ NO_3^- = 412.9$ mhos.

We require the equivalent conductance of $0.0020F$ HCl. Clearly, we would wish to remove from arithmetical involvement the K^+ and the NO_3^- ions. We can do this by adding equations b and c and, from the result obtained, subtracting equation a. This leaves us solely with the H^+ and Cl^- ions that constitute the chemical identity of the solute in the HCl solution. Therefore

(b) $\qquad \lambda 0.0020N\ K^+ + \lambda 0.0020N\ Cl^- = 145.8$ mhos
(c) $\qquad \lambda 0.0020N\ H^+ + \lambda 0.0020N\ NO_3^- = 412.9$ mhos

---

$(b + c):\ \lambda 0.0020N\ K^+ + \lambda 0.0020N\ NO_3^- + \lambda 0.0020N\ H^+$
$\qquad\qquad\qquad\qquad\qquad + \lambda 0.0020N\ Cl^- = 558.7$ mhos
(a) $\qquad \lambda 0.0020N\ K^+ + \lambda 0.0020N\ NO_3^- = 140.5$ mhos

---

$(b + c - a):\ \lambda 0.0020N\ H^+ + \lambda 0.0020N\ Cl^- = 418.2$ mhos.

Consequently, the required equivalent conductance of solution is

$$\Lambda 0.0020F\ HCl = 418.2\ \text{reciprocal ohms}$$

■ PROBLEM 2: To derive the limiting equivalent conductance (infinite dilution) of a weak electrolyte.

Calculate the limiting equivalent conductance of the weak acid, $HC_2H_3O_2$, from the following data:

$\Lambda_0 NaC_2H_3O_2 = 90.6$ mhos; $\Lambda_0 NaCl = 125.3$ mhos; $\Lambda_0 HCl = 423.5$ mhos.

Solution. The procedure here is identical with that of the previous example, although the point to be stressed here is the necessary independence presumed to be individually and fully enjoyed by one equivalent of each of the weak electrolyte's ions at infinite dilution; that is, its limiting equivalent conductance. This may be ascertained from the limiting equivalent conductances of the ions of strong electrolytes. It is obvious that, in this instance, we need to exclude arithmetically the Na^+ and Cl^- ions of the salts that have been provided for this purpose. Therefore, by adding

$$[\Lambda_0 NaC_2H_3O_2] = \lambda_0 Na^+ + \lambda_0 C_2H_3O^- = 90.6 \text{ mhos}$$

and

$$[\Lambda_0 HCl] = \lambda_0 H^+ + \lambda_0 Cl^- = 423.5 \text{ mhos},$$

we obtain

$$\lambda_0 Na^+ + \lambda_0 H^+ + \lambda_0 C_2H_3O^- + \lambda_0 Cl^- = 514.1 \text{ mhos},$$

from which we subtract

$$\lambda_0 Na^+ + \lambda_0 Cl^- = 125.3 \text{ mhos}$$

and obtain

$$\lambda_0 H^+ + \lambda_0 C_2H_3O_2^- = 388.8 \text{ mhos}.$$

In thus constituting the necessary expression for a solution of acetic acid, our result implies a 100% ionization of the weak acid at its limiting equivalent conductance of infinite dilution; therefore

$$\Lambda_0 HC_2H_3O_2 = 388.8 \text{ reciprocal ohms}.$$

■ PROBLEM 3: To derive the percentage of ionization of a weak electrolyte from its specific conductance at a given finite dilution, and of its equivalent conductance at infinite dilution.

At $t°C$, the specific conductance of a 0.010 F NH_4OH solution ($= 0.010N$ $NH_4OH = 0.010N$ NH_3 aqueous) is 9.63×10^{-5} reciprocal ohm (mho). At infinite dilution, the equivalent conductance of the electrolyte is evaluated as 235 reciprocal ohms. Calculate therefrom the percentage of ionization of this weak base.

Solution. The degree of ionization of the weakly ionized solute is readily obtained as the quotient of:

$$\left[\frac{\text{equivalent conductance at finite dilution (\textit{limited} ionization), } \Lambda}{\text{equivalent conductance at infinite dilution (\textit{complete} ionization), } \Lambda_0} \right]$$

Hence, our first task is to convert the given specific conductance to equivalent conductance. We recall that the specific conductance represents the conductance of a cube one centimeter on edge (and thus, sensibly, the volume of one milliliter), and that the equivalent conductance defines the conductance of one entire equivalent weight of the solute. With respect to the given normality, it will take 100,000 ml (i.e., 1000/0.010) to provide an entire equivalent weight of the solute to conform to the given conductance. That is,

equivalent conductance, $\Lambda 0.010N$ NH$_4$OH

$$= \text{specific conductance} \times \frac{1000}{\text{normality}}$$

$$= 9.63 \times 10^{-5} \text{ mho} \times \frac{1000}{0.010 \text{ equivalent}}$$

$$= 9.63 \text{ mhos/equivalent.}$$

We may now derive the desired percentage of ionization by substituting in the relationship initially provided:

$$\frac{\Lambda 0.010N \text{ NH}_4\text{OH}}{\Lambda_0 \text{NH}_4\text{OH}} \times 100 = \% \text{ ionization of } 0.010F \text{ NH}_4\text{OH}$$

$$= \frac{9.63 \text{ mhos}}{235 \text{ mhos}} \times 100 = 0.041\% \text{ ionized}$$

■ PROBLEM 4:   To determine the degree of ionization of a weak electrolyte when the limiting equivalent conductances of its cation and anion are known.

Calculate the degree of ionization of a $0.10N$ HC$_2$H$_3$O$_2$ solution having an equivalent conductance of 5.21 reciprocal ohms, and given the following additional data: λ_0H$^+$ = 348 reciprocal ohms; λ_0 C$_2$H$_3$O$_2$ = 40.8 reciprocal ohms.

Solution. In conformity with Kohlrausch's law, the limiting equivalent conductance, Λ_0, of any electrolyte solute is the sum of the equivalent conductances of its anion(s) and cation(s) in infinite dilution, whereat the independent migration of such ions is presumed to be unimpaired by electrical forces of restraint. Applying the given data, we obtain

$$\Lambda_0 \text{HC}_2\text{H}_3\text{O}_2 = \lambda_0\text{H}^+ + \lambda_0\text{C}_2\text{H}_3\text{O}_2^-$$

$$= 348 \text{ mhos} + 40.8 \text{ mhos}$$

$$= 388.8 \text{ mhos.}$$

Thereupon, for the degree of ionization (α) —

$$\alpha = \frac{\Lambda 0.10N \text{ HC}_2\text{H}_3\text{O}_2}{\Lambda_0 \text{HC}_2\text{H}_3\text{O}_2}$$

$$= \frac{5.21 \text{ mhos}}{388.8 \text{ mhos}} = 0.0134, \text{ as sought.}$$

■ PROBLEM 5: To evaluate the solubility product constant (K_{sp}) of a difficultly soluble solute from pertinent measurements of conductance.

A saturated aqueous solution of the sparingly soluble, unhydrolyzed salt of the generalized divalent cation, M^{2+}," and the divalent anion "Z^{2-}" has a specific conductance of 1.80×10^{-6} reciprocal ohm. The specific conductance of pure water is determined to be 4.50×10^{-8} reciprocal ohm. At infinite dilution, the values of the limiting equivalent conductances of the ions are as follows:

$$M^{2+} = 88 \text{ reciprocal ohms,}$$
$$Z^{2-} = 95 \text{ reciprocal ohms.}$$

Calculate from these data the value of the solubility product constant, K_{sp} MZ.

Solution. Mass action for the slight solubilization of the solute,

$$MZ_{(s)} \rightleftarrows M^{2+} + Z^{2-}$$

is expressed by

$$[M^{2+}][Z^{2-}] = K_{sp}MZ.$$

Required, then, are the molar concentrations of M^{2+} and Z^{2-} ions. These we may obtain by prior progressive evaluations of Λ_{MZ} and of the solubility of the salt MZ in terms of normality, utilizing the expression

$$\Lambda_{MZ} = \kappa_{MZ} \times \frac{1000}{N_{MZ} \text{ solution}}$$

wherein κ designates specific conductance.

The given information affords a ready determination of the specific conductance of the solute MZ, conforming to the derivation

$$\kappa\ MZ = \kappa \text{ Sat'd. solution } MZ - \kappa\ H_2O$$
$$= 1.80 \times 10^{-6} - 4.50 \times 10^{-8}$$
$$= 1.76 \times 10^{-6} \text{ mho.}$$

For a sparingly soluble solute

$$\Lambda_{solute} \approx \Lambda_0 solute$$

(as developed in the descriptive principles of this chapter) the value of Λ_{MZ} is also readily available because it is the sum of the mobilities of the component ion; that is

$$\Lambda_{MZ} \approx \Lambda_0\ MZ = \lambda_{0M^{+2}} + \lambda_{0Z^{2-}}$$
$$= \quad 88 \ + \ 95$$
$$= 183 \text{ mhos.}$$

By substitution in the conductance-concentration expression we may now obtain the normality of the saturated solution; thus

$$183 = 1.76 \times 10^{-6} \times \frac{1000}{N}$$

whereupon

$$N = \frac{1.76 \times 10^{-3}}{183} = 9.6 \times 10^{-6} \, (= 9.6 \times 10^{-6} \text{ g-equiv } MZ/\text{liter}).$$

By definition, a $1N$ solution of a salt supplies one gram-equivalent weight of each of its ions to the liter. In this instance of a di-divalent salt, it is represented by one-half of the gram-formula weight of the salt; therefore, for a $9.6 \times 10^{-6}N$ solution the corresponding molarity for the concentration will be

$$9.6 \times 10^{-6}/2 = 4.8 \times 10^{-6}M \text{ in dissolved } MZ.$$

As $4.8 \times 10^{-6}M$ MZ must dissociate to yield $4.8 \times 10^{-6}M$ M^{2+} ion and $4.8 \times 10^{-6}M$ Z^{2-} ion, we now have the respective required molar concentrations of ions to substitute in the pertinent expression for mass action that evaluates the desired solubility product constant, K_{sp}

$$[4.8 \times 10^{-6}][4.8 \times 10^{-6}] = K_{spMZ}$$
$$K_{spMZ} = 2.3 \times 10^{-11}, \textit{ as sought.}$$

■ PROBLEM 6:   To determine the relationship of total measured resistance in a cell to the specific conductance of an electrolyte.

A conductance cell is filled with $0.015N$ NaCl solution at a temperature for which the specific conductance of this electrolyte measures 0.00326 reciprocal ohm. The total resistance of the cell is found to be 255 ohms. When, however, at the same temperature the identical cell is filled, instead, with a solution of a certain unknown electrolyte, the total resistance of the cell declines to 145 ohms. Calculate from this information the specific conductance of the unknown electrolyte.

Solution.   We first establish the relationship between the total measured resistance of the cell and the specific resistance. As observed, the resistance of any conductor varies directly with the distance between electrolytes, or length of conductor, and inversely with the area of cross-section; hence

measured total resistance

$$= \text{specific resistance} \times \left[\frac{\text{distance between electrodes, cm}}{\text{area of cross-section, cm}^2} \right].$$

The ratio of the distance between electrodes to the area of cross section is itself a constant for the particular cell, no matter what electrolyte is placed therein. Consequently, once this ratio is established for the solution of a known electrolyte (in this instance, the $0.10N$ NaCl), it will apply unalterably to the unknown solution. The data of the problem provide directly the measured total resistance of the NaCl solution for substitution in the initial relationship by which it has been defined; and inasmuch as the likewise-required specific resistance therein is the reciprocal of the specific conductance (see data provided for problem) we have

$$\text{specific resistance, NaCl} = \frac{1}{0.00326 \text{ mho}}$$

$$= 306.6 \text{ ohms.}$$

We may now obtain via substitution

$$\left[\frac{\text{distance between electrodes, cm}}{\text{area of cross section, cm}^2}\right] = \frac{\text{measured resistance, ohm-cm}}{\text{specific resistance, ohm-cm}}$$

$$= \frac{255 \text{ ohm-cm}}{306.6 \text{ ohm-cm}}$$

$$= 0.832.$$

This constant for the cell now applies with equal validity when the unknown electrolyte solution is introduced. Therefore we may obtain the specific resistance of the unknown by once again substituting appropriately in the initial expression:

$$\text{specific resistance of ``unknown''} = \frac{\text{measured resistance}}{0.832}$$

$$= \frac{145 \text{ ohms}}{0.832}$$

$$= 174 \text{ ohms.}$$

Consequently, as

$$\text{specific conductance} = \frac{1}{\text{specific resistance}}$$

we derive

$$\text{specific conductance of unknown} = \frac{1}{174 \text{ ohms}},$$

$$= 0.00575 \text{ reciprocal ohm, } as\ sought.$$

EXERCISES

1. At $t°C$, a given solution of a certain weak electrolyte is 3.61% ionized and reveals a specific resistance to electrical conductivity of 2077 ohms. The equivalent

conductance measures 7.04 reciprocal ohms. Calculate for this electrolyte solution:

(a) specific conductance.

(b) concentration normality.

(c) equivalent conductance at infinite dilution.

2. Given, at $t°C$, the following equivalent conductances at infinite dilution:

$$H^+ = 346 \text{ reciprocal ohms}$$

$$Q^- = 48 \text{ reciprocal ohms.}$$

Calculate therefrom the percentage ionization of a solution of $0.070F$ HQ (a, hypothetical monoprotic weak acid) which possesses a specific conductance, of 3.55×10^{-4} reciprocal ohms.

3. Calculate for a solution of HX (a hypothetical monoprotic strong acid) which at $t°C$, provides one full gram-equivalent weight of solute in 40.0 liters of solution and which has an equivalent conductance of 404 reciprocal ohms:

(a) normality.

(b) specific conductance.

(c) specific resistance at the determined normality.

(d) percentage of apparent ionization at the determined normality if, at infinite dilution (zero concentration), the equivalent conductance is evaluated as 553 reciprocal ohms.

4. Given the following data at $t°C$:

$$\Lambda_0 HNO_3 = 419 \text{ mhos}$$

$$\Lambda_0 NH_4C_2H_3O_2 = 115 \text{ mhos}$$

$$\Lambda_0 NH_4NO_3 = 147 \text{ mhos}$$

calculate therefrom the value of $\Lambda_0 HC_2H_3O_2$.

5. At $t°C$, a conductance cell containing $0.100N$ KCl solution yields a measured cell resistance of 70.5 ohms. The specific conductance of this electrolyte is 0.0129 reciprocal ohm. When the same cell is filled, instead, with electrolyte solution, Q, the total resistance of the cell measures 138 ohms. Calculate from the foregoing:

(a) *constant* of the cell that evaluates the ratio:

$$\left[\frac{\text{distance between electrodes (in cm)}}{\text{area of cross-section of conductor (in cm}^2)} \right].$$

(b) specific resistance of electrolyte solution "Q."

(c) specific conductance of electrolyte solution "Q."

6. The specific resistance of an electrolyte in an electrolytic cell measures 18.5 ohm cm. The inert electrodes being used are 290 square centimeters in cross-section and are placed at a distance of 15 cm from each other. Calculate the total resistance (in ohms) of the electrolyte to passage of the current.

7. The equivalent conductance of a $0.0200N$ HY solution (the monoprotic weak acid of the hypothetical anion Y^-) at the presumed complete ionization of its

solute at infinite dilution (Λ_0) is 730 reciprocal ohms. The specific resistance of the solution at given normality is 1187 ohms. From this information calculate for the solution of HY:

(a) specific conductance.

(b) percentage of ionization of its solute.

(c) molar concentrations provided of ions of H$^+$ and of Y^-.

CHAPTER 5

COLLIGATIVE PROPERTIES OF IONIC SOLUTIONS

It will be noted that *conductance*, as explained in the previous chapter, has not been characterized as a colligative property. The term *colligative* (from the Latin, *colligatus*, meaning *joined together*) denotes the intimate relationships of the properties of solutions in terms of total numbers of all particles present, both with and without electrical charges. As the electrical conductivity of a solution is a function exclusively of the charged particles therein (ions), in a strictly definitive sense we would be necessarily excluding from consideration the electrically uncharged molecules that are always present in a dissolved weak electrolyte, be it acid, or base, or pseudo-salt, because these do not actually transport current.

REVELATIONS OF PARTICLE NUMBERS —RAOULT'S LAW

The colligative properties of solutions are roughly intermediate between the physical characteristics of their component solutes and solvents, with due allowances made for their respective concentrations. Consequently, the evaluations of the colligative effects of a solution correspond approximately to the arithmetical average (sometimes complicated by the factor of concentration) of the particular specific physical property of the solute(s) and solvent. The fact that an exact average is rarely realized bespeaks the nonideality of most solutions, for the "ideality" of a solution presumes a purely physical intermixture of solute and solvent without the slightest chemical interaction. This latter condition hardly ever prevails, although ideality is manifestly most closely approached as the solution becomes more and more dilute with respect to the dissolved solute, and particularly so when the solute is a nonelectrolyte. In such cases, the molecules of neutral solute are little affected by the *van der Waals* forces of attraction to molecules of solvent.

The observation of properites that commonly initiates any theoretical investigation of the specific colligative behavior of a solution was first made

86

by F. M. Raoult (France, 1881), who generalized that the vapor pressure of each volatile component of a solution is directly proportional to the *mole fraction* that represents the concentration of the components. This generalization is conveniently expressed in any of the following equivalent ways for the dilute solution of a nonelectrolytic nonvolatile solute (sugar, for example) dissolved in a volatile solvent (water, for example):

1. The vapor pressure of the solution is directly proportional to the mole fraction of solvent therein.

2. The depression of the vapor pressure of the volatile solvent, when the solute dissolves in it, is directly proportional to the mole fraction of solute.

3. The depression of the vapor pressure of the volatile solvent in a solution is directly proportional to the concentration of molality in the solution.

In accordance with the implications of Raoult's law, it follows that the lowered freezing point and elevated boiling point of a volatile solvent in solution (experimentally manifested colligative effects) result directly from the depression of the vapor pressure of the solvent (at a constant temperature) by the dissolved solute, and in proportion to the mole fraction of solute present. These alterations are unrelated to the specific chemical identities of the particles of solute. They depend solely upon the mere physical presence of the particles, and they are irrespective, as well, of whether the particles be exclusively ions, exclusively molecules, or any variable combinations of both. The greater the over-all number of particles of solute in a given weight of solvent, the greater the magnitude of change. The lower, that is, becomes the freezing point, and the higher the boiling point of the solution, as contrasted with the respective physical constants of the pure solvent itself.

For a gram-formula weight (molecular weight) of un-ionized solute dissolved in 1000 g of solvent H_2O (that is, a solution of 1 molal concentration), provided that the solute exerts no significant vapor pressure of its own, the boiling point of the resultant aqueous solution at one atmosphere of pressure is very close to 100.52°C. This is an elevation of 0.52°C over the normal boiling temperature, 100.00°C, of the pure solvent. The freezing point of the same 1 *m* solution is − 1.86°C; that is, 1.86°C below the normal freezing temperature, 0.00°C, of the pure solvent.

APPLICATIONS OF RAOULT'S LAW

The constancies of the changes in the physical responses of a solvent (to a given set of conditions) which permit the familiar, ready calculations of the molecular weights of un-ionized, nonvolatile solutes, are conveniently applied as algebraic modifications of Raoult's law in mathematically determining the colligative effects of electrolyte solutes (or "ionogens") upon the physical constants of the solvent.

These may be expressed as follows:

$$m \times K_{vp} = \Delta P_{vp}$$

$$m \times K_{fp} = \Delta T_{fp}$$

$$m \times K_{bp} = \Delta T_{bp}$$

where m = Total molality of all species (molecular, ionic)

K_{vp} = Constant for depression of molal vapor pressure
(depends on temperature for $1m$ aqueous solutions)

K_{fp} = Constant for depression of molal freezing point
($1.86°C$ for $1m$ aqueous solutions)

K_{bp} = Constant for elevation of molal boiling point
($0.52°C$ for $1m$ aqueous solutions)

Δ = Change in pressure (P) or temperature (T) from respective
normal values of pure solvents

Our immediate interests here are the applications of the foregoing representations of Raoult's law in determining the degree of ionization of weak electrolytes, and even the apparent degrees of dissociation of strong electrolytes. With respect to this latter category should be recalled the "clustering" of ions already present, which are nonetheless effectively impaired in their mobility by the mutual electrical "drags" of their opposite charges.

Clearly, the value of m (molality) for equilibria of electrolytes cannot be construed as the usual gram-formula weight of solute in 1000 g of H_2O. Definition must be enlarged to encompass the increased total of free particles that has resulted from ionization or dissociation, by virtue of which the vapor pressure, and freezing point, of a solution of nonvolatile ionogens are always lower, and the boiling point is always higher than those respectively of a solution of a nonvolatile nonelectrolyte of similar composition, with respect to weight of solvent and number of gram-formula weights of the dissolved parent solute.

As NaCl is completely ionic and provides actually one mole each of Na^+ and Cl^-, for a total of $2 \times 6.02 \times 10^{23}$ particles, we should expect that in any aqueous solution containing the same weight of solvent one gram-formula weight of NaCl would depress the freezing point and elevate the boiling point of the solvent twice as much as would a gram-formula weight of the non-electrolyte $C_6H_{12}O_6$ (glucose).

Because of their completely ionic identities, solutions of $1.0m$ K_2SO_4, $1.0m$ $K_3[Fe(CN)_6]$, and $1.0m$ $K_4[Fe(CN)_6]$ would supply, respectively, $3 \times 6.02 \times 10^{23}$, $4 \times 6.02 \times 10^{23}$ and $5 \times 6.02 \times 10^{23}$ ions and consequently yield effects three, four, and five times, respectively, those of the glucose solution. It must be assumed that comparisons are being made for similar gram-formula weights of all the parent solutes in identical volumes of the pure solvent.

For example, the experimental observation that the freezing point of a $1.0m$ NaCl solution $-3.72°C$ — is somewhat larger than the theoretically predicted one — that is, not quite the "expected" full depression, $2 \times -1.86°C$ actually, the freezing point is $-3.30°C$) merely strengthens the concept of a certain extent of ion-pairing in response to electrical attractions between Na^+ and Cl^- ions, and the consequent reduction in the number of free and completely independent ions to something less than $2 \times 6.02 \times 10^{23}$. As the molality of the solute increases, so does the extent of divergence (on the basis of complete dissociation) from the theoretical value calculated without benefit of corrections for activity.

What amounts to the same thing, as concentrations of electrolyte increase, disproportionately greater weights of parent solute are required to provide additional particles to compensate for the increased losses of free ions that result from increased clustering or pairing, and thus to depress freezing points and raise boiling points, in direct proportion to the changes in the total molality of the solution. Table 5:1 shows some convenient appraisals of the foregoing considerations. The corresponding values of ΔT_{fp} should be compared numerically with the five-fold and ten-fold increases in molality of (b) and (c), respectively, over that of (a); and with the two-fold relationship of (c) to (b) as well.

TABLE 5:1

ΔT_{fp} — Freezing Point Depressions of Some Electrolytes

Conc'n of parent solute in gram-formula weights/ 1000 g of H_2O	2 ions mono-monovalent NaCl	2 ions mono-monovalent KNO_3	3 ions mono-divalent K_2SO_4	2 ions di-divalent $MgSO_4$	4 ions mono-trivalent $K_3[Fe(CN)_6]$
(a) 0.01	0.03606	0.03587	0.05010	0.0300	0.06260
(b) 0.05	0.1758	0.1719	0.2280	0.1294	0.2800
(c) 0.10	0.3470	0.3331	0.4319	0.2420	0.5370

What has been expressed in the foregoing data with respect to the colligative effects of ionogens upon the freezing points of solutions, applies with equal force to alterations in vapor pressures and boiling points. Experimental evaluations of the percentages of ionization of solutes by determining the boiling points of solutions are, however, necessarily limited in sensitivity, primarily because of the far narrower ranges afforded for measurements. The elevation of boiling temperature of 1000 g of H_2O to a gram-formula weight of solute is only $0.52°C$, and, illustratively, at $25°C$ the depression in molal vapor pressure is only 0.425 mm. Compare these values with the depression of $1.86°C$ in molal freezing point.

MOLE NUMBER — THE VAN'T HOFF FACTOR

Our forthcoming mathematical delineations of extent of ionization of electrolytes by the colligative properties of solutions will make use of the concept of "mole number." This factor expresses a measure of the degree of deviation of an electrolyte solution from the standard colligative effects presumed in an idealistic application of Raoult's law. Mole number, designated by the letter i, and often called the "van't Hoff factor," is mathematically defined as follows:

i (mole number)

$$= \frac{\text{experimentally observed change in freezing point}}{\text{calculated change in freezing point assuming no ionization}}$$

$$= \frac{\Delta T_{fp} \text{ electrolyte at } m \text{ molality}}{\Delta T_{fp} \text{ non-electrolyte at } m \text{ molality}}$$

$$= \frac{\Delta T_{fp}}{m_s \times K_{fp}}$$

(where m_s is molality at zero ionization).

Clearly, as i represents the *equilibrium total number of* particles, both ions and molecules, derived from a single formula unit of a solute, its value for dilute solutions will range from 1 for nonelectrolytes (zero ionization) to 2, for completely ionic mono-monovalent and di-divalent electrolytes, such as KCl and $MgSO_4$, respectively; 3, for completely ionic mono-divalent and di-monovalent electrolytes, such as K_2SO_4 and $MgCl_2$, respectively; and 4, for a completely ionic mono-trivalent type, such as $K_3[Fe(CN)_6]$. In actuality, the value of i approaches these limiting whole numbers in acceptably close proximity only in very dilute solutions where interionic attractions are at a minimum. As described in connection with the activity concept, the repellent forces between ions of like charge at excessively high concentrations are strong enough to cause their activities to exceed their actual concentrations. Consequently, experimental readings at extremely high concentrations may lead to mole numbers greater than those of the limiting values assigned to i in dilute solutions.

Weakly ionized electrolytes, such as acetic acid or ammonia, would yield values of i only slightly larger than one. And, as expected, deviations from limiting values of i become greater for strong electrolytes of similar molality (identical gram-formula weights of parent solute in a fixed weight of solvent) as the valence charges upon the ions increase. These effects may be observed in Table 5:2.

TABLE 5:2

Derivation of Mole Number from Depression of Freezing Point

Electrolyte	Concentration in dissolved gram-formula weights to 1000 g H_2O				
	0.005	0.010	0.050	0.100	0.500
KCl	1.96	1.94	1.89	1.86	1.80
NaCl	1.95	1.94	1.90	1.87	1.81
H_2SO_4	2.59	2.47	2.21	2.17	1.99
K_2SO_4	2.86	2.80	2.57	2.46	2.32
$MgCl_2$	2.81	2.72	2.68	2.66	2.90
$CuSO_4$	1.55	1.45	1.22	1.12	0.93
$MgSO_4$	1.62	1.57	1.30	1.21	1.07
$K_3[Fe(CN)_6]$	3.51	3.31	3.02	2.85	2.45

APPLICATIONS OF DATA ON FREEZING POINT TO DEGREE OF IONIZATION

1. *When Cations and Anions Are Equal in Numbers*

The extents of ionization of mono-monovalent weak electrolytes, e.g., HCN and $HC_2H_3O_2$, as well as the apparent degrees of ionization of strong electrolytes of both mono-monovalent and di-divalent types, e.g., KCl and $MgSO_4$, respectively, may be deduced as follows:

let α = degree of dissociation or ionization
and m_s = molality of solute at zero ionization.

Accordingly, for any of the named electrolytes that must ionize or dissociate by the generalized reaction,

$$AB \rightleftarrows A^+ + B^- \qquad or \qquad CD \rightleftarrows C^{2+} + D^{2-},$$

the molalities of the respective ions at equilibrium are, each, $m_s\alpha$; and the molal concentration of un-ionized solute — or, with strong electrolytes, the clusters of ions that are equivalent practically to un-ionized molecules — $m_s - m_s\alpha$.

The notation m_s, again, represents molality of solute at zero ionization; the symbol α denotes the actual degree of ionization. The total number of moles of all species, irrespective of chemical identity, then is additively for either $AB \rightleftharpoons A^+ + B^-$ or $CD \rightleftharpoons C^{2+} + D^{2-}$:

$$(m_s - m_s\alpha) + m_s\alpha + m_s\alpha, \qquad \text{or} \qquad m_s(1 + \alpha).$$

This may now be substituted in the mathematical expression of Raoult's law that defines change in freezing point, namely:

$$m_s \times K_{fp} = \Delta T_{fp},$$

where m_s refers to an ideal nonelectrolyte, which for an electrolyte must be subsitituted by the total number of moles of all species at equilibrium, molecules, or ion clusters, and free ions, positive and negative. Substituting the $m_s(1 + \alpha)$ of the electrolyte for the total m yields for the aqueous system,

$$m_s(1 + \alpha) \times 1.86 = \Delta T_{fp}$$

which rearranges to the form

$$\alpha = \frac{\Delta T_{fp}}{1.86 \times m_s} - 1.$$

This condensed formula is serviceable directly with all electrolytes that provide equal numbers of both cations and anions.

2. *When Cations and Anions Are Not Equal in Numbers*

The degrees of ionization in this category are those of "apparent" nature that, in essence, interpret the extent of ion-clustering or ion-pairing of strong electrolytes caused by pronounced interionic attractions. These electrolytes are of types $(A^+)_2D^{2-}$ and $C^{2+}(B^-)_2$, illustrated by K_2SO_4 and $MgCl_2$, respectively, and of even the higher polyvalent character, as represented by $(Al^{3+})_2(SO_4^{2-})_3$, $(K^+)_3[Fe(CN)_6^{3-}]$, and $(K^+)_4[Fe(CN)_6^{4-}]$.

Approaching the evaluations of α in precisely the same manner as before, we can obtain the following condensed expressions for ready use:

(a) For salts like $MgCl_2$ and/or K_2SO_4: in 1000 g of H_2O

$$\begin{aligned}
\text{total moles of particles} &= m_s - m_s\alpha + 2m_s\alpha + m_s\alpha \\
&= m_s + 2m_s\alpha \\
&= m_s(1 + 2\alpha)
\end{aligned}$$

whence, from the Raoult formulation of $m_s \times 1.86 \times \Delta T_{fp}$, we deduce, by substituting the total molality of the electrolyte for m_s, that

$$\alpha = \frac{1}{2}\left(\frac{\Delta T_{fp}}{1.86 \times m_s} - 1\right).$$

(b) For a salt like $K_3[Fe(CN)_6]$ we would obtain:

$$\alpha = \frac{1}{3}\left(\frac{\Delta T_{fp}}{1.86 \times m_s} - 1\right).$$

(c) For the salt of the type of $K_4[Fe(CN)_6]$,

$$\alpha = \frac{1}{4}\left(\frac{\Delta T_{fp}}{1.86 \times m_s} - 1\right).$$

(d) Derivation of α for a tri-divalent salt such as $Al_2(SO_4)_3$ would conform as follows —

$$\text{total moles of particles in 1000 g } H_2O = m_s - m_s\alpha + 2m_s\alpha + 3m_s\alpha$$
$$= m_s + 4m_s\alpha$$
$$= m_s(1 + 4\alpha)$$

whence, again by substitution:

$$\alpha = \frac{1}{4}\left(\frac{\Delta T_{fp}}{1.86 \times m_s} - 1\right).$$

Consistently then, for any electrolyte yielding the variable n number of ions for a unit of the formula of parent solute by presumed total ionization or dissociation, α is merely a fractional multiple of

$$\left(\frac{\Delta T_{fp}}{1.86 \times m_s} - 1\right), \text{ namely: } \alpha = \frac{1}{n-1}\left(\frac{\Delta T_{fp}}{1.86 \times m_s} - 1\right).$$

APPLICATIONS OF DATA ON BOILING POINT AND VAPOR PRESSURE TO DEGREE OF IONIZATION

Delineations of α by elevation of boiling point or drop in vapor pressure (when these methods are experimentally feasible) lead to the same condensed relationships as derived from data on freezing point. All that is necessary is merely the substitution of ΔT_{bp} or ΔP_{vp} for ΔT_{fp}, and the substitution of the constant 1.86 (the K_{fp}) by 0.52 (the K_{fp}), or by the numerical value of K_{vp} at the specific temperature of investigation, as applicable.

INTERCONVERTIBILITY OF MOLALITY AND MOLE NUMBER OF ELECTROLYTE

The common link between total molality of the species of an electrolyte at equilibrium and the mole number (the van't Hoff factor, i) of its solution is seen as ΔT_{fp} or ΔT_{bp} or ΔT_p — depending upon whether the *modus operandi* of the experiment has been by alterations in freezing point, boiling point, or vapor pressure, respectively. Illustratively, relationships conforming to the arithmetical definitions already given are

$$\text{total molality (all particles)} \times K_{fp} = \Delta T_{fp}$$

and

$$i \times K_{fp} \times m_s = \Delta T_{fp}$$

which yield upon combination:

$$m_{\text{(total electrolyte)}} \times K_{fp} = \Delta T_{fp} = i \times K_{fp} \times m_s$$

which reduces to

$$m_{\text{(total electrolyte)}} = i \times m_s$$

Knowing any two of these terms, we may then determine the third.

Sample Calculations

■ PROBLEM 1: To determine concentrations of ions and the extent of ionization of a weak electrolyte whence they derive, employing for such purposes the freezing point of the given solution.

Calculate the molal concentrations of ions and the degree of ionization of a 0.0500 molal (m) solution of $HC_2H_3O_2$, given its freezing point as $-0.0947°C$.

Solution. The total number of particles of solute that must be present to depress the freezing point of 1000 g of H_2O to the stated extent, is determined as follows:

1 mole of particles produces a depression of 1.86°C (by definition)

$\dfrac{1}{1.86}$ mole of particles in total produces a depression of 1.00°C

$\left(\dfrac{1}{1.86} \times 0.0947\right)$ mole of particles in total produces a depression of 0.0947°C.

These particles comprise *both* un-ionized molecules and ions produced by ionization, empirically represented by

$$HC_2H_3O_2 \rightleftarrows H^+ + C_2H_3O_2^-$$

If we denote by x the number of gram-formula weights of the initially introduced $HC_2H_3O_2$ (0.0500 mole/1000 g of H_2O) that undergo ionization, the numbers of moles of each species of solute present in the solution at equilibrium are:

$$(0.0500 - x) = \text{moles of } HC_2H_3O_2 \text{ remaining un-ionized}$$
$$x = \text{moles of } H^+ \text{ ion produced by ionization}$$
$$x = \text{moles of } C_2H_3O_2^- \text{ ion produced by ionization}$$

for a total number of particles, regardless of chemical identity, of

$$(0.0500 - x) + x + x$$

or, $0.0500 + x$ particles.

We have thus established the requisites for a valid mathematical equation; namely

$$\text{total moles of particles} = \text{total moles of particles}$$

wherein

$$0.0500 + x = \frac{1}{1.86} \times 0.0947$$

whence

$$x = 0.0509 - 0.0500$$
$$= 0.00090 \text{ mole of ionized } HC_2H_3O_2$$

and, consequently, of H^+ and also $C_2H_3O_2^-$ ions, and all in the total weight of 1000 g of the solvent H_2O.

We likewise have:

$$0.0500 - 0.00090 = 0.0491 \text{ mole}/1000 \text{ g of } H_2O,$$

of un-ionized $HC_2H_3O_2$

The degree of ionization, α, of $HC_2H_3O_2$ is the ratio of the number of moles of the parent solute ionized to the number of moles thereof initially introduced into solution. Therefore

$$\alpha = \frac{0.00090}{0.0500} = 0.018 = 1.8\% \text{ of ionization.}$$

Our answers are then, collectively:

$$H^+ = 9.0 \times 10^{-4}m;$$
$$C_2H_3O_2^- = 9.0 \times 10^{-4}m;$$
$$HC_2H_3O_2 = 0.0491m;$$
$$\alpha = 0.018.$$

Had the concentration of the original solution been expressed as $0.0500F$, rather than $0.0500m$ — representative, that is, of the very same number of gram-formula weights of the solute but dissolved in 1000 ml of solution — the numerical values of the answers would still have held without significant change; for at this dilution of a weak electrolyte,

$$0.0500m \approx 0.0500F.$$

We might have elected to use the condensed formula that was developed and presented in the section dealing with descriptive principles in this chapter. Our work, then, would have been facilitated; thus, ionization would check out as

$$\alpha = \frac{1}{n-1} \times \left(\frac{1.86 m_s}{\Delta T_f} - 1\right)$$

$$\alpha = \frac{1}{1} \times \left(\frac{0.0947}{1.86 \times 0.0500} - 1\right) = \frac{1.018 - 1}{1} = 0.018 (= 1.8\%), \text{ as before;}$$

and the molal concentration of each ion would then be the arithmetical product of α times m_s, leading to the same results as already calculated for the pertinent species.

■ PROBLEM 2: To evaluate from the freezing point of its solution the apparent degree of ionization of a strong electrolyte of valence type $(A^+)_2 B^{2-}$, and the concentrations of species it offers.

Calculate the apparent percentage of ionization (dissociation) of a 0.100 m solution of K_2SO_4, given its freezing point as $-0.457°C$. Determine also the concentrations of *ions* and *ion-clusters*.

Solution. A total of 1 mole of particles in 1000 g of H_2O produces a depression of $1.86°C$ (by definition); consequently, a total of $\frac{1}{1.86}$ mole produces a depression of $1.00°C$, and

$$\left(\frac{1}{1.86} \times 0.457\right)$$

mole of particles, in total, will be required to depress the freezing point by $0.457°C$.

The equation for complete dissociation of K_2SO_4 is

$$K_2SO_4 \rightarrow 2K^+ + SO_4^{2-}.$$

If we set

x = number of moles of solute apparently dissociated,

then

$(0.100 - x)$ = number of moles of ion-clusters (aggregating $2K^+$, SO_4^{2-})
$2x$ = moles of free K^+ ion
x = moles of free SO_4^{2-} ion

for a total number of moles of all particles, regardless of chemical identity of

$$(0.100 - x) + 2x + x \, (= 0.100 + 2x).$$

Hence, we may equate

$$0.100 + 2x = \frac{0.457}{1.86}.$$

from which

$$2x = 0.245 - 0.100$$

and

$$x = 0.073 \text{ mole of } K_2SO_4 \text{ apparently dissociated.}$$

It follows then, that

free K^+ ion

$$= 2x = 2 \times 0.073 = 0.146 \text{ mole}/1000 \text{ g } H_2O = 0.146m$$

free SO_4^{2-} ion

$$= x = 0.073 \text{ mole}/1000 \text{ g } H_2O = 0.73m$$

and

ion-clusters (aggregating $2K^+$, SO_4^{2-})

$$= (0.100 - 0.073) = 0.027m.$$

The apparent percentage of dissociation of the given $0.100m$ solution of K_2SO_4 is, therefore

$$\frac{0.073}{0.100} \times 100 = 73\%.$$

Our alternative use of the condensed formula,

$$\alpha = \frac{1}{n-1} \times \left(\frac{\Delta T_f}{1.86 m_s} - 1\right)$$

would likewise have produced directly:

$$\alpha = \frac{1}{2}\left(\frac{0.457}{1.86 \times 0.100} - 1\right)$$

$$\frac{2.46 - 1}{2}$$

$$= 0.73, \text{ or } 73\% \text{ as before.}$$

This result signifies that 27% of the original solute remains in the form of ion-clusters (or ion-aggregates).

■ PROBLEM 3: To derive concentrations of electrolyte and degree of ionization from mole number.

Calculate the concentrations of species derivable from the solute in a $0.00500\ m$ solution of formic acid, a weak electrolyte ($HCHO_2$), and the extent of ionization thereof, given the mole number (i) of the solution as 1.014.

Solution. If $x =$ the number of moles of $HCHO_2$ that ionize according to the equation

$$HCHO_2 \rightleftarrows H^+ + CHO_2^-,$$

then

$$x = \text{molality of } H^+ \text{ ion}$$
$$x = \text{molality of } CHO_2^- \text{ ion}$$

and

$$(0.0050 - x) = \text{molality of un-ionized } HCHO_2.$$

The total number of moles to 1000 g of H_2O of all species present, regardless of their specific chemical identities, is thus

$$(0.0050 - x) + x + x = (0.0050 + x) \text{ total moles.}$$

By general definition,

$$\text{electrolyte } m \times K_{fp} = \Delta T_f$$

and

$$i \text{ (the mole number)} = \frac{\Delta T_f}{K_{fp} \times m_s}$$

Combining these two equations yields

$$\text{electrolyte } m \times K_{fp} = \Delta T_f = i \times K_{fp} \times m_s$$

and, for our purposes,

$$\text{electrolyte } m \times K_{fp} = i \times K_{fp} \times m_s$$

Substitution within this relationship yields

$$(0.00500 + x) \times 1.86 = 1.014 \times 1.86 \times 0.00500$$

whence, with the value 1.86 cancelling from both sides,

$$0.00500 + x = 0.005070$$
$$x = 0.000070 \text{ mole } HCHO_2/1000 \text{ g } H_2O, \text{ ionized.}$$

Therefore,

$$H^+ \text{ ion concentration} = 7.0 \times 10^{-5}m$$
$$CHO_2^- \text{ ion concentration} = 7.0 \times 10^{-5}m$$

and

$$\text{un-ionized } HCHO_2 \text{ concentration} = (0.00500 - 0.000070)$$
$$= 0.00493 \text{ mole}/1000 \text{ g } H_2O.$$

Required, still, is the extent of ionization. This develops as follows:

$$\alpha \text{ (degree of ionization)} = \frac{7.0 \times 10^{-5}}{0.00500}$$

$$= 0.014; \text{ or } 1.4\% \text{ ionization.}$$

Collecting our several results

$$H^+ + 7.0 \times 10^{-5} \, m;$$
$$CHO_2^- = 7.0 \times 10^{-5} \, m;$$
$$\text{un-ionized } HCHO_2 = 4.9 \times 10^{-3} \, m;$$
$$\alpha = 0.014.$$

■ PROBLEM 4: To derive concentrations of species at equilibrium and mole number of a strong electrolyte from its apparent degree of dissociation.

Calculate the concentrations of species at equilibrium and the value of i (mole number) for a $0.100m$ solution of $K_3Fe(CN)_6$, at a temperature for which the apparent dissociation of the salt is 65.0%.

Solution. The salt dissociates as follows:

$$K_3Fe(CN)_6 \rightarrow 3K^+ + Fe(CN)_6^{3-}$$

As $\alpha = 0.65$ (apparent degree of ionization), the concentration of each of the pertinent species present is:

$$K^+ = 3m_s\alpha = (3 \times 0.100 \times 0.650) = 0.195m$$
$$Fe(CN)_6^{3-} = m_s\alpha = (0.100 \times 0.650) = 0.065m$$

and

$$\text{ion-clusters (aggregating } 3K^+, Fe(CN)_6^{3-}) = (m_s - m_s\alpha)$$
$$= (0.100 - 0.065) = 0.035m.$$

The total moles of all species to 1000 g of solvent H_2O is $(m + 3m\alpha)$ = 0.295; that is, the electrolyte solution is effectively 0.295 molal. By definition,

$$\text{total molality of electrolyte} = \frac{\Delta T_f}{1.86}$$

and

$$i \times m_s = \frac{\Delta T_f}{1.86}.$$

Hence

$$i \times m_s = \text{total molality of electrolyte}$$

or

$$i = \frac{\text{total molality of electrolyte}}{m_s},$$

from which substitution leads to

$$i = \frac{0.295}{0.100} = 2.95.$$

Collating our results, we have:

$$K^+ = 0.195m$$
$$Fe(CN)_6^{3-} = 0.065m$$
$$\text{Cluster aggregates } (3K^+, Fe(CN)_6^{3-}) = 0.035m;$$
$$i = 2.95$$

■ PROBLEM 5:   To evaluate the concentration of a strong electrolyte from the freezing point and apparent ionization of its solution.

A solution of $Ca(NO_3)_2$ freezes at $-0.413°C$, and exhibits a conductivity that indicates an apparent ionization of 64.7%. Calculate therefrom with respect to the parent solute,

(a) the molality, m, of each of the particles in solution derived from the solute.

(b) the formality of the concentration of the parent solute, assuming that for this dilute solution, $m \approx F$.

Solution.

(a) The solution of $Ca(NO_3)_2$ yields its individually independent ions and its ion-clusters (or aggregates, Ca^{2+}, $2NO_3^-$) in accordance with the stoichiometry,

$$Ca(NO_3)_2 \rightarrow Ca^{2+} + 2NO_3^-$$

If $x =$ the number of moles of $Ca(NO_3)_2$ dissolved in 1000 g of solvent H_2O, then, because the solute is 64.7% apparently ionized

$$(x - 0.647x) = \frac{\text{the molality of the}}{\text{ion-aggregates of } Ca(NO_3)_2}$$

$$0.647x = \text{the molality of the } Ca^{2+} \text{ ion}$$

$$2 \times 0.647x = \text{the molality of the } NO_3^- \text{ ion.}$$

Consequently, the total molality of the solution of all species of solute, irrespective of their individual chemical identities, is

$$(x - 0.647x) + 0.647x + 2(0.647x).$$

But, total molality of solution (m) is also expressed as

$$m \times K_{fp} = \Delta T_f$$

which, for the given aqueous solution, is numerically

$$m = \frac{0.413°C}{1.86°C}.$$

Hence, we may solve for the value of x in the equation

$$(x - 0.647x) + 0.647x + 2(0.647x) = \frac{0.413}{1.86},$$

whereupon we obtain

$$x + 1.294x = \frac{0.413}{1.86}$$

$$x = \frac{0.413}{2.294 \times 1.86} = 0.0968 \text{ mole/1000 g } H_2O.$$

Therefore, the molal concentration of each of the different species of solute becomes:

$$Ca^{2+} = 0.647 \times 0.0968 = 0.0626m$$
$$NO_3^- = 2(0.647 \times 0.0968) = 0.125m$$
$$Ca(NO_3)_2 \text{ ion-aggregates} = 0.0968 - (0.647 \times 0.0968)$$
$$= 0.0342m.$$

(b) As it is being assumed for this dilute solution that

molality of species $\approx$ molarity of species

it follows directly that the apparent ionization of the parent solute forms 0.0626 mole/liter of Ca^{2+} ion (as obtained in part a of this example). In conformity with the stoichiometry of solubilization, plainly the molar concentration of the $Ca(NO_3)_2$ dissociated must equal the molar concentration of the Ca^{2+} ion; the former is then, likewise, $0.0626M$. Substituting within the following equation:

$$\left[\frac{\text{moles/liter of } Ca(NO_3)_2 \text{ dissociated}}{\text{total gram-formula weights/liter of } Ca(NO_3)_2 \text{ dissolved } (= F)}\right] \times 100$$
$$= \% \text{ dissociation,}$$

we obtain

$$\frac{0.0626}{F} \times 100 = 64.7$$

$$F = \frac{0.0626}{0.647} = 0.0967F \; Ca(NO_3)_2, \text{ as sought.}$$

The formality of the concentration thus obtained conforms to the interpreted data of the problem. As it is given that only 64.7% of the solubilized solute, $Ca(NO_3)_2$, is dissociated into free ions, plainly 35.3% must exist as ion-aggregates of the form $[Ca^{2+}, 2NO_3^-]$. Multiplying the calculated formality of concentration of 0.0967 (= 0.0967 mole/liter of the parent solute, calcium nitrate) by 0.353 (the extent of ion-clustering) confirms directly the value of the concentration of such ion-aggregates. This has already been determined, in part a, to be $0.0342m$ ($\approx 0.0342M$). The solution of the entire problem could have been approached from the point of view of ion-clustering just as validly as from that of apparent dissociation.

————————

■ PROBLEM 6: To determine the freezing point of a solution when molarity, density, and percentage of ionization are known.

A $6.00F$ solution of $HC_2H_3O_2$ at $t°C$ has a density of 1.046 g/ml, and is ionized 0.180%. Calculate therefrom the freezing point of the solution. Given constant for molal freezing point (K_{fp}) for $H_2O = 1.86°C$ to the mole.

Solution. Our first concern is the transposing of concentration from molarity to molality, as required for any valid comparative appraisals of the colligative properties of a solution. The weight of solute that is provided by 1000 ml of the given solution is

$$6.00 \text{ g-formulas } HC_2H_3O_2 \times \frac{60.0 \text{ g}}{1 \text{ g-formula}}$$

$$= 360 \text{ g of } HC_2H_3 \text{ solute}/1000 \text{ ml of solution.}$$

With the density of the solution given as 1.046 g/ml, the 360 g of $HC_2H_3O_2$ solute will be contained in a total weight of solution of

$$1000 \text{ ml} \times 1.046 \text{ g/ml} = 1046 \text{ g of solution (solute + solvent).}$$

Therefore,

$$\text{weight of solvent} = \text{weight of solution} - \text{weight of solute}$$
$$= 1046 \text{ g} - 360 \text{ g}$$
$$= 686 \text{ g of } H_2O.$$

Were the $HC_2H_3O_2$ completely un-ionized, its molality (as represented by the number of moles of it present in 1000 g of H_2O) would have to be

$$\frac{6.00 \text{ g-formulas } HC_2H_3O_2}{686 \text{ g } H_2O} \times 1000 = 8.75m \text{ } HC_2H_3O_2,$$

on the assumption that ionization is nil.

Note, in connection with this value, the extremely large error that would have been committed had molarity ($6.00 \text{ } M$) been inadvertently (or deliberately, for the sake of arithmetical expediency) accepted as molality; the error would be $(8.75 - 6.00)/8.75$, or approximately 31%.

Ionization however, is not nil; and as it does proceed, it must contribute to the actual total molality of the over-all solution. This will be the sum, plainly, of the individual molalities of all the species of solute present — ions and molecules alike, and without regard for chemical identity. We derive, then, from the actual 0.18% of ionization undergone by the parent solute ($8.75m$) the following:

$$HC_2H_3O_2 \rightleftharpoons H^+$$

molalities: $(8.75 - [8.75 \times 0.00180])$ (8.75×0.00180)
          remaining un-ionized

$$+ \quad C_2H_3O_2^-$$
$$(8.75 \times 0.00180).$$

Their total sum (all species) yields

molality of solution $= (8.75 - [8.75 \times 0.00180]) + (8.75 \times 0.00180)$
$$+ (8.75 \times 0.00180)$$
$$= 8.75 + (8.75 \times 0.00180)$$
$$= 8.77m.$$

We may now obtain the theoretical freezing point depression of the solution by

$$\Delta T_f = m \times K_{fp}$$

whence

$$\Delta T_f = 8.77 \times 1.86$$
$= 16.3°C$ the theoretical depression of the freezing point of the solvent H_2O of the solution, from initial freezing temperature of the *pure* solvent, $0.00°C$.

Theoretically, therefore, the actual freezing point of the solution is $-16.3°C$.

EXERCISES

1. Calculate the percentage of apparent ion-pairing of the pure solute in a solution containing 0.0250 g-formula weight of the fully ionic salt KCl in 250 g of H_2O, which freezes at $-0.362°C$.

2. Calculate the percentage of apparent ion-clustering of the pure solute in a solution containing 0.05000 g-formula weight of Na_2SO_4 in 125 g of H_2O, which freezes at $-1.93°C$.

3. Calculate the formality of the concentration of a solution of $MgCl_2$ that freezes at $-0.423°C$, and which exhibits an apparent dissociation of 65.0%.

4. Calculate for a given $0.250F$ solution of $Ca(NO_3)_2$ having an apparent ion-clustering of 12.0%
   (a) total molality
   (b) freezing point
   (c) boiling point.

5. The freezing point of an 80-ml sample of a pure aqueous solution containing a total dissolved quantity of 0.0090 mole of the monoprotic acid, H"T" (identity concealed) is given as $-0.225°C$. Calculate the percentage of ionization of this acid. Assume the density of the solvent at the temperature of measurement to be 1.0 g/ml.

6. A $0.215m$ solution of the tri-monovalent pseudo-salt, MA_3 (hypothetical) freezes at $-0.5650°C$. Calculate therefrom
   (a) total collective molality of all species of solute in solution

(b) number of moles in 1000 g of H_2O of each of the individual species of solute present in the solution at equilibrium

(c) the percentage of dissociation of the pseudo-salt.

7. Assuming an idealized complete absence of interionic attractions, calculate the weight, in grams, of K_2SO_4 that when solubilized in 250 g of H_2O yields a solution that freezes at the same temperature as a $0.180\ m$ solution of KNO_3 (likewise idealized).

8. Calculate for $0.0400m$ solution of formic acid ($HCHO_2$ a monoprotic acid) at a temperature for which the van't Hoff mole number (i) of the solute is 1.028

(a) freezing point of the solution

(b) elevation of the boiling point of the solution

(c) total collective molality of all species of solute in the solution at equilibrium

(d) percentage of ionization of the solute.

9. At a temperature of $t°C$, a $0.650m$ solution of $NaHSO_4$ undergoes an essentially complete dissociation into Na^+ and HSO_4^- ions, but the monoprotic anion, HSO_4^-, undergoes thereafter 12.6% of ionization into H^+ and SO_4^{2-} ions. Calculate with respect to this solution

(a) molality of each of the different species of solute present in the solution at equilibrium

(b) freezing point of the solution (assuming ionization remains unchanged)

(c) boiling point of the solution (assuming ionization remains unchanged)

(d) van't Hoff mole number (i) of the dissolved compound.

10. An aqueous $0.62m$ solution of the monoprotic weak acid, HA (derivative of the hypothetical anion A^-) freezes at $-1.198°C$. If, at its boiling point, the solute is 2.6% more highly ionized than at its freezing point, what is

(a) percentage of ionization of the acid at its boiling point?

(b) boiling point of the solution?

CHAPTER 6

THE CHEMICAL FORMULA AND ITS ARITHMETIC

SIGNIFICANCE OF THE FORMULA

Let us start by interpreting as many as possible of the points of information that a formula contrives to register in terms of weight or volume, remembering that every formula is in itself a revelation of the laws of quantitative composition. And, as such, its representation of a specific substance must rest solidly upon an experimental foundation. What, then, does a formula tell us about experimental facts? We enumerate the more obvious.

1. *The elemental qualitative composition of the compound.*

In H_2O, for example, we observe a qualitative make-up of hydrogen and oxygen. The symbols of the elements — supplied in every table of atomic weights — become habitually familiar with constant use, and should be associated with the names of the elements themselves — because in nearly all instances, they are either the first letter or the initial and a significant letter of the name.

2. *The relative proportions of the whole numbers of atoms constituting the make-up of the formula unit of the substance.*

In the absence of any specific information or familiarity with the substance, nothing can be inferred as to the exact numbers of atoms comprising the unit of the formula. Thus, the ratio of atoms of hydrogen to atoms of oxygen in the liquid state of "H_2O" — water — most simply depicted as $2:1$, may also be expressed as multiples; namely, $4:2$, $6:3$, $8:4$, etc. Clearly, the formula unit cannot be validly represented as a "molecule" unless a molecular weight as determined by experiment has delimited the weight of the formula unit and permitted the assignment of the exact numbers of the atoms present. Were a molecular weight to be established here as approximately 18 amu, or 18 g/mole (as would be

105

true with this compound as a *vapor*), a molecular assignment of H_2O would be inescapable; for the sum of the weights of all the atoms in this formula corresponds to this value.

Were the molecular weight to be established, instead, as some multiple, n, of 18 such as 36 or 54 or 72, which are common in the liquid and solid states of this compound, the identities of the corresponding formula units would be, respectively:

$$H_4O_2[= (H_2O)_2], \qquad H_6O_3[= (H_2O)_3], \qquad H_8O_4[= (H_2O)_4].$$

Empirical and molecular formulas. What is being stressed here is the terminological differentiation between *molecular* formula and *empirical* formula. If, in the absence of a molecular weight, assignable through experiment, we cannot correctly ascribe a molecular identity to a formula unit, we use the very simplest relationship that expresses a whole-number ratio of atoms — that is, the empirical formula. As in the given instance of H_2O vapor, the empirical formula may prove to be the molecular formula as well.

It must not be supposed from what has preceded that the difference between empirical formula and molecular formula becomes purely academic, because the chemical differences revealed between H_2O as a vapor and $(H_2O)_n$ as liquid or solid are relatively minor. Physical differences are actually profound.

The chemical properties of a substance are hardly subverted by whatever assignment is made, correctly or incorrectly. *Rust* remains rust whether it be designated as Fe_2O_3, or $(Fe_2O_3)_n$, or something else altogether. The coincidental similarities of chemical properties of molecular forms as compared to empirical forms must, of necessity, be restricted to the interconversions of physical states (gas, liquid, solid) of one and the same compound.

We must not assume, for example, that because CH_2O is the empirical formula of the specific compound *glucose*, this same empirical formula may not actually also express either the empirical or the molecular formula of an entirely different compound. And, indeed, it does — *formaldehyde* for one, which is both empirically and molecularly CH_2O.

The molecular formula of glucose, however, is established as $C_6H_{12}O_6$, on the basis of a molecular weight experimentally determined to be approximately 180. Following in the same vein, neither should it be assumed that once a true molecular formula is established for a given compound that this cannot be duplicated for an entirely different compound. Thus, $C_6H_{12}O_6$ is the molecular formula of not only glucose (grape sugar), but also of other simple sugars, fructose, mannose, and galactose. These are examples of *isomerism* — the molecular manifestations of different compounds of identical molecular formulas. Isomers, which are responsible for much of the varying chemistry of carbon compounds, and which contribute to the large number of the latter, owe their differences to the dissimilarities of the architectural or structural arrangements of the comprising atoms.

In most instances, the appearance of a formula that, by means of a common numerical divisor, can be reduced further to whole numbers of atoms, stamps such a formula as molecular rather than empirical. This is always true with gases. In other cases, judicious caution is to be exercised, particularly, in the solid state wherein atomic arrangements reflect giant aggregations of individual units whose restricted mobilities are responsible for crystal or lattice structures. Thus, the dimeric formula unit, Hg_2Cl_2, used for solid mercuric chloride, may not validly be called a "molecule," because it is divisible by 2 to yield the monomer, HgCl (a classic but now-discarded formulation of the compound). Nevertheless, structural arrangements of the atoms within it make preferable and more revealing its empirical designation as the dimer.

3. *The relative weight ratios of the combining elements.*

This responds to the simple addition of the atomic weights of all of the constituent atoms of the elements of the compound. Illustratively, for the compound, potassium dichromate ($K_2Cr_2O_7$), the weight ratios are

$$(2 \times 39.102) : (2 \times 51.996) : (7 \times 15.9994)$$
$$\text{potassium} \quad \text{chromium} \quad \quad \text{oxygen}$$

Clearly, weight ratios are completely independent of whether the formula is molecular or empirical; for, the given ratio is only one of an infinite number of equivalencies obtainable by multiplying or dividing the specific ratios of total atomic weights of each of the individual comprising elements by any common number, integral or decimal.

4. *The weight of the formula unit, molecular weight or ionic weight as definable by experiment.*

This must be the sum of all of the atoms comprising the particular formula unit. Thus, for the $K_2Cr_2O_7$ just employed, we obtain

$$
\begin{aligned}
\text{potassium} \ (2 \times 39.102) &= 78.204 \\
\text{chromium} \ (2 \times 51.996) &= 103.992 \\
\text{oxygen} \quad (7 \times 15.9994) &= \underline{111.996} \\
K_2Cr_2O_7 \ \text{formula weight} &= 294.192
\end{aligned}
$$

On a microscopic scale, this signifies that each single formula unit of the compound has a mass of 294.192 amu; and on the equivalent macroscopic scale, 1 mole ($= 6.02 \times 10^{23}$ formula units of $K_2Cr_2O_7$) has a gram-formula weight of 294.192 g. That is, one gram-formula weight of $K_2Cr_2O_7$, 294.192 g, provides of the individual elements:

2 g-atoms of K ($2 \times 6.02 \times 10^{23}$ atoms), weighing 78.204 g

2 g-atoms of Cr ($2 \times 6.02 \times 10^{23}$ atoms, weighing 103.992 g

7 g-atoms of O ($7 \times 6.02 \times 10^{23}$ atoms) weighing, 111.996 g.

Had we realistically expressed the exact constitution of this compound as two potassium ions ($2K^+$) and one dichromate ion ($Cr_2O_7^{2-}$), electrostatically or coulombically mutually associated, the following relationships would clearly apply:

$$\text{potassium ion in } K_2Cr_2O_7 \begin{cases} \text{ionic weight } K^+, 78.204 \text{ amu } K^+/ \\ \text{formula unit } K_2Cr_2O_7 \\ \text{gram-ionic weight } K^+, 78.204 \text{ g} \\ K^+/\text{gram-formula (mole) } K_2Cr_2O_7 \end{cases}$$

$$\text{dichromate ion in } K_2Cr_2O_7 \begin{cases} \text{ionic weight } Cr_2O_7^{2-}, 215.988 \text{ amu} \\ Cr_2O_7^{2-}/\text{formula unit } K_2Cr_2O_7 \\ \text{gram-ionic weight } Cr_2O_7^{2-}, 215.988 \text{ g} \\ Cr_2O_7^{2-}/\text{gram-formula (mole) } K_2Cr_2O_7. \end{cases}$$

5. *The volume, at STP, of any compound in its gaseous state.*

By definition, the gram-molecular volume (GMV) of a gas is the volume that 6.02×10^{23} molecules of it occupy at 0°C and 760 mm of pressure, or standard temperature and pressure (STP). This has been experimentally ascertained as virtually 22.4 liters, regardless of the chemical identity of the gas. Additionally, the gram-molecular weight of a gas (GMW) represents the mass of this Avogadro number of molecules. It is inescapable, consequently, that one gram-formula weight of each and every gas at STP may be interchangeably and equivalently designated either as a volume — *ideally* constant at 22.4 liters — or as a weight — its gram molecular weight.

PERCENTAGE COMPOSITION FROM FORMULA

We are interested here in ascertaining the percentage-by-weight of each element present in any given substance. We use sodium chromate (Na_2CrO_4) by way of illustration. As percentage-weight composition signifies the parts-by-weight of any individual constituent present in one hundred parts by weight of the entire independent unit in which that constituent appears, clearly, our first task is to establish the total weight of the entire independent unit; that is, its formula-weight, in amu/unit or grams/mole. Hence, from atomic weights

$$
\begin{aligned}
\text{sodium} &= 2Na = (2 \times 22.9898) = 45.9796 \text{ g} \\
\text{chromium} &= 1Cr = (1 \times 51.996) = 51.996 \text{ g} \\
\text{oxygen} &= 4O = (4 \times 15.9994) = 63.9976 \text{ g}
\end{aligned}
$$

$$Na_2CrO_4 \text{ gram-formula weight} = 161.973 \text{ g.}$$

Consequently,

$$\text{percentage of sodium} = \frac{45.9796 \text{ g Na/mole Na}_2\text{CrO}_4}{161.973 \text{ g Na}_2\text{CrO}_4/\text{mole Na}_2\text{CrO}_4} \times 100$$

$$= 28.387\%.$$

$$\text{percentage of chromium} = \frac{51.996 \text{ g Cr/mole Na}_2\text{CrO}_4}{161.973 \text{ g Na}_2\text{CrO}_4/\text{mole Na}_2\text{CrO}_4} \times 100$$

$$= 32.102\%.$$

$$\text{percentage of oxygen} = \frac{63.9976 \text{ g O/mole Na}_2\text{CrO}_4}{161.973 \text{ g Na}_2\text{CrO}_4/\text{mole Na}_2\text{CrO}_4} \times 100$$

$$= 39.511\%.$$

A check proves the correctness of computations; for,

$$28.387\% + 32.102\% + 39.511\% = 100.000\% \text{ sample.}$$

It is readily apparent that we might have spared ourselves the necessity of working out the computation for *one* of these components merely by subtracting the sum of the other two percentages from 100. Any inaccuracy in either of these two, however, or in both, inevitably leads to an error in the third as well. Consequently, it is highly desirable to work out all the components individually and, thereafter, to check upon accuracy by ascertaining that the sum of all component percentages totals 100 (with deference to any slight deviations in the last decimal place, reflective of "rounding off" adjustments).

It is to be noted that percentage composition remains completely unaffected by any differentiations we might make with respect to true molecular formula and its empiricized version. As the former is always an integral multiple of the latter, clearly the arithmetical factor that mutually relates the two would appear in both numerator and denominator of the percentage-composition ratio, and thus would cancel out.

SYNTHESIS OF FORMULA FROM PERCENTAGE COMPOSITION

How do we build up a formula from the experimental data of percentage-weight composition? Merely to say that "formula synthesis" is just the reverse of "formula analysis" might well prove to be a case of misplaced confidence in taking for granted the systematic arithmetical procedures that we seek to establish, with which correct results may be facilitated. Inasmuch as we have already used Na_2CrO_4 in illustrating formula analysis, we use it again in formula synthesis. The systematic procedure remains the same no matter which of the many hundreds of thousands of substances we employ in illustration.

We start with the following experimental data.

Elemental Analysis of Sample of Substance

Weight of sample taken for analysis	43.1264 g
Percentage of sodium in sample	28.387%
Percentage of chromium in sample	32.102%
Percentage of oxygen in sample	39.511%.

Before proceeding, we take note that the actual weight of the sample taken, aside from the critically important exactitude of its value, has no significance other than that it represents a convenient weight responsive to the sensitivity of laboratory apparatus and handling. The larger the weight of the sample, the smaller (as a rule) the percentage of error in measurement of weight. In any event, we stress that even though percentage composition signifies parts-by-weight of component to 100 g of the sample, we need not by any means take 100 g of the sample. This is inherent, clearly, in the "fractional weight ratio times 100" that collectively defines percentage composition.

The formula applicable to the given sample of compound can be established according to the arithmetical sequence that follows.

1. *Transpose the percentage composition of each component to its respective weight in the sample.*

Thus

$$\text{sodium: } 28.387\% \text{ of } 43.1264 \text{ g} = 0.2837 \times 43.1264$$
$$= 12.2423 \text{ g Na}$$

$$\text{chromium: } 32.102\% \text{ of } 43.1264 \text{ g} = 0.32102 \times 43.1264$$
$$= 13.8444 \text{ g Cr}$$

$$\text{oxygen: } 39.511\% \text{ of } 43.1264 \text{ g} = 0.39511 \times 43.1264$$
$$= 17.0397 \text{ g O.}$$

Check: 12.2423 g + 13.8444 g + 17.0397 g = 43.1264 g of sample.

It should be pointed out here that the weights in grams of the constituent elements, which we have just derived, are precisely what the chemist has actually determined experimentally in his analysis of the sample in the laboratory. In this particular instance arithmetical transposition of these weights to percentage composition was completely unnecessary inasmuch as we promptly converted them back again to designations as grams. In thus appearing to "put the cart before the horse" we have but pointed up the necessary recognition of the *constancy* of percentage composition of all samples of any pure substance and, consequently, of the advantage of defining the sample in these exact terms (percentage composition) under all conditions. Actual weights (grams) of the individual constituents have, on the other hand, as many different numerical values as there are different weights of sample taken for analysis.

2. *Convert the weights of the individual components to their corresponding numbers of gram-atoms.*

Inasmuch as one gram-atom of each element is represented by its atomic weight in grams, any given weight of an element may be converted to numbers of gram-atoms merely by dividing by the atomic weight; that is

$$\frac{\text{grams}}{\text{grams/gram-atom}} = \text{gram-atoms}$$

Therefore

$$^{Na}\left[\frac{12.2423}{22.9898}\right] {}^{Cr}\left[\frac{13.8444}{51.996}\right] {}^{O}\left[\frac{17.0397}{15.9994}\right]; \textit{numerical quantities in grams}$$

$$= {}^{Na}[0.5325] \, {}^{Cr}[0.266] \, {}^{O}[1.0650]; \textit{ numerical quantities in gram-atoms.}$$

This provisional formula may now be transcribed to the conventional whole-atom form of the empirical formula simply by dividing each of its numerical values by the smallest of them. In this instance the common divisor is 0.266, and the formula sought for the compound, *sodium chromate*, which conforms to the given experimental data, checks out as Na_2CrO_4.

It may be noted that the respective divisions for Na and O did not yield exact integers — the trivial deviations therefrom resulting from the uncertainty of the last significant decimal of each. This situation is normally to be expected in calculations of this type, since not only because such figures have been rounded off in prior arithmetic but also because, even at best, the laboratory weighing operations basically responsible for the entire procedure are always slightly imperfect. Consequently, judicious allowance must be made for the very slight discrepancies of experimental or mathematical insensitivities.

On the other hand, it is readily possible for a very highly divergent fractional remainder to persist even after the initial division by the smallest of the fractional quantities in the provisional formula. Illustratively, an analysis of a sample of ferric oxide has provided the following data:

weight of sample taken	90.3858 g
weight of iron in sample	63.2188 g
weight of oxygen in sample	27.1670 g.

Consequently, the provisional formula is

$$^{Fe}\left[\frac{63.2188}{55.847}\right] {}^{O}\left[\frac{27.1670}{15.9994}\right]; \textit{numerical quantities in grams}$$

$$= {}^{Fe}[1.132] \, {}^{O}[1.6980]; \textit{ numerical quantities in gram-atoms}$$

whence, division by the smaller, 1.132, continues to yield a decimal remainder; thus,

$^{Fe}[1.000]^O[1.500]$; *numerical quantities in gram-atoms.*

The empirical formula the smallest integral multiple of this. Multiplying by 2, we obtain

Fe_2O_3, *as the empirical formula.*

MOLECULAR FORMULA FROM EMPIRICAL FORMULA

We have already established the simple whole-number relationship of the molecular formula to the empirical formula. Consequently, once the empirical formula has been established in accordance with percentage-composition procedure it may be converted to the specific multiple thereof (1, 2, 3, etc.) that identifies the true molecular identity of the substance. The specific multiple is the value of the ratio

$$\frac{\text{gram-molecular weight}}{\text{gram-empirical weight}}.$$

The gram-empirical weight is simply the sum of the atomic weights of all atoms in the empirical formula; the gram-molecular weight is the collective mass of 6.02×10^{23} units of the substance, derived by experiment. Molecular weights may be determined in the laboratory in many ways. The one to which mathematical implications is confined here is that of determining the gram-molecular volume of a gas, 22.4 liters at standard $0°C$ and 760 mm of Hg pressure. Moreover, as liquids may frequently be vaporized without chemical decomposition, the implications of gaseous GMV may, under such conditions, be translated into terms applicable to them as well. Elsewhere, we deal with determining the molecular weight of solids, by the alterations they effect in the freezing point, boiling point, and vapors pressure of the solvents in which they are dissolved.

We illustrate the derivation of a molecular formula by using a simple hydrocarbon, ethane. The following experimental data are available:

density of ethane $= 1.342$ g/liter at STP
percentage by weight of carbon $\qquad$ 79.888%
percentage by weight of hydrogen $\qquad$ 20.112%.

Derivation of the compound's empirical formula and empirical weight conforms to the following mathematical sequence:

$$\text{Provisional formula} = {}^C\left[\frac{79.888}{12.01115}\right] {}^H\left[\frac{20.112}{1.00797}\right]$$
$$^C[6.651]\ ^H[19.953]$$
$$^C\left[\frac{6.651}{6.651}\right] {}^H\left[\frac{19.953}{6.651}\right]$$

Empirical formula $= CH_3$; and, consequently, for this formula

Empirical weight $= (1 \times 12.01115) + (3 \times 1.00797)$

total weight of   total weight of
carbon   hydrogen

$$= 15.03506 \text{ g/g-formula ethane.}$$

The gram-molecular weight of a gas is always the weight of its gram-molecular volume; therefore, employing the stated density at STP,

$$GMW = 1.342 \frac{g}{liter} \times 22.4 \frac{liters}{GMV}$$

$$= 30.061 \frac{g}{GMV}.$$

Clearly, the factor that transforms gram-empirical formula weight of ethane to gram-molecular weight is

$$\frac{30.061}{15.035} = 2.$$

Again, the slight deviations from an exact integer, normally encountered, is well within acceptable limits of experimental sensitivities of determining densities and weight percentages. In any event, the necessarily exact whole number is always conspicuously revealed, in the given instance, 2. Therefore, the molecular formula of ethane, as corresponds to its GMW, is C_2H_6.

VALENCE — COMBINING CAPACITIES OF ATOMS

The implications of atomic theory and of definite proportions by weight are that atoms in chemical union form chemical bonds that are specific both in numbers and in types for each identifiable substance. The implications of multiple-weight proportions are that the same atoms in chemical union may, under different sets of conditions, form different numbers and types of bonds, provided that altogether different compounds result. It is not to be inferred from either of these observations that only one type of chemical bond may appear in the same compound; but, the number of each type of bond is specific and fixed.

The number of bonds that an atom, or group of atoms in a collective unit, can form is called its *valence*. Valences numerically describe, consequently, the chemical combining capacities of atomic species and assist in the writing of chemical formulas by correctly reflecting the proportions of the combining weights of the associated atoms. Although valences for a particular atom (or polyatomic group) may be of a multiple character, as a rule the different compounds in which the atom appears under ordinary or conventional conditions reveals it as having just one or two different valences. These we might characterize as the "common valences."

The chemical principle that the total valence of one atom must equal the total valence of any other atom with which it combines, provides an immediate and useful theoretical device for the correct writing of chemical formulas. When this is coupled with a systematic nomenclature, it ensures that all considerations of the weight of the respective components are correctly revealed.

The standard of valence is the hydrogen atom, with an accepted valence of 1. Consequently, chemical combination of a single atom of any other element "E," with one atom of hydrogen, necessitates the same valence, inasmuch as the bond H–E is unavoidably common to both elements. Hence, E must likewise have a valence of 1 in this combination of atoms. Had E combined with two atoms of hydrogen, H_2E, E would be revealing a valence of 2. If the compound formed were H_3E, the valence of E would have been stated as 3.

As some elements do not yield stable binary compounds with hydrogen (two different elements), their valences require elucidations from the bonds that they form with other elements whose valences are deducible from hydrogen-atom associations in binary compounds. Thus, in HCl, the chlorine shows a valence of 1; and, therefore, the generalized atom E is in the different compounds, ECl, ECl_2, ECl_3, and ECl_4, show valences of 1, 2, 3, and 4, respectively. Were E to have combined with oxygen in the different compounds, E_2O, EO, E_2O_3, and EO_2, the element E, would have revealed respective valences of 1, 2, 3, and 4, on the logical premise that the compound water, H_2O, shows that its oxygen atom has a valence of 2. We thus stress not only the chemical principle of equality of total valence of all the atoms of each component element in a binary compound, but also the convenient applicability of the mathematical precept, "things equal to the same thing are equal to each other," to the determination of formulas.

A helpful device in making valence assignments requisite to correct formulations is to associate with the particular atom (or polyatomic group, acting as a collective unit) its conventional appearance, wherever possible, in a compound in which it is bonded with hydrogen atom(s). This permits, usefully, assignments of valences to polyatomic groups as they appear in ternary compounds (comprised of three different elements). We illustrate by constructing the formulas of various compounds from the symbols of their pertinent atomic constituents and reconciling the valences of the latter in various reference compounds in which they appear:

A. Symbols of elements and polyatomic groups (also called radicals) pertinent to our illustrations:

K (potassium)	S (sulfur)
Ca (calcium)	OH (hydroxyl radical)
Al (aluminum)	NO_3 (nitrate radical)
NH_4 (ammonium radical)	SO_4 (sulfate radical)
	PO_4 (phosphate radical)

B. Reference compounds in which appear the various species given in *A* preceding:

HCl, HNO_3, H_2SO_4, H_2S, H_3PO_4, KCl, $CaCl_2$, $AlCl_3$, NH_4Cl, HOH (H_2O)

With the information that has been provided, it should prove a simple matter to construct the formulas of all of the compounds that the species in the left-hand column of *A* form with those in the right-hand column, utilizing for the purpose the pertinent valences of each species as shown in the reference compounds of *B*. These are, by inspection,

1 for K	1 for NH_4	1 for OH_3
2 for Ca	2 for S	2 for SO_4
3 for Al	1 for OH	3 for PO_3.

Therefore, we may write for the required compounds

K_2S, potassium sulfide	CaS, calcium sulfide
KOH, potassium hydroxide	$Ca(OH)_2$, calcium hydroxide
KNO_3, potassium nitrate	$Ca(NO_3)_2$, calcium nitrate
K_2SO_4, potassium sulfate	$CaSO_4$, calcium sulfate
K_3PO_4, potassium phosphate	$Ca_3(PO_4)_2$, calcium phosphate
Al_2S_3, aluminum sulfide	$(NH_4)_2S$, ammonium sulfide
$Al(OH)_3$, aluminum hydroxide	NH_4OH, ammonium hydroxide
$Al(NO_3)_3$, aluminum nitrate	NH_4NO_3, ammonium nitrate
$Al_2(SO_4)_3$, aluminum sulfate	$(NH_4)_2SO_4$, ammonium sulfate
$AlPO_4$, aluminum phosphate	$(NH_4)_3PO_4$, ammonium phosphate.

NOMENCLATURE

In most instances, the names of compounds prove to be systematic in character. Here and there familiar "trivial" or proprietary historical names still linger. Compounds are named by mentioning the more metallic element first; thereafter, if the compound is binary, the suffix *ide* is appended to the second (less metallic) element after dropping its last syllable. Hence, $MgBr_2$ is referred to as *magnesium bromide;* CdS is *cadmium sulfide;* $AlCl_3$ is *aluminum chloride*.

The inevitable exceptions, largely due to historical associations, include the names of binary acids, that is, compounds, in which hydrogen atom(s) are combined with a single different element. In such cases, the acidic character of aqueous solutions of the dissolved compounds is given special recognition, in deference to their importance in chemical work, by contracting the word "hydrogen" to its root *hydro* and then appending *ic* to the root of the second element. Thus, for example,

HF   is correctly both hydrogen fluoride gas and hydrofluoric acid
HCl   —   hydrogen chloride   —   hydrochloric acid
HBr   —   hydrogen bromide   —   hydrobromic acid
HI   —   hydrogen iodide   —   hydriodic   acid
H_2S —   hydrogen sulfide   —   hydrosulfuric acid
H_2Se —   hydrogen selenide   —   hydroselenic acid
H_2Te —   hydrogen telluride   —   hydrotelluric acid.

As already observed, time-honored proprietary names persist despite all attempts to systematize nomenclature. Thus, gaseous NH_3 — systematized as *nitrogen hydride* or *hydrogen nitride* (H_3N) — remains familiarly *ammonia*. And *water* (H_2O) is hardly ever called *hydrogen oxide*, although it would certainly not be incorrect to do so.

When compounds are ternary, and contain oxygen, the particular suffix to be appended depends upon the relative numbers of atoms of oxygen that are bonded to the central atom. Thus *-ate*, as a suffix, designates a greater number of bonded oxygen atoms than does the suffix *-ite*; consequently, $KClO_3$ is *potassium chlorate*, and $KClO_2$ is *potassium chlorite*. As KClO and $KClO_4$ also exist, it is obvious that additional terminological refinements are in order. To indicate the smaller content of oxygen in KClO compared to $KClO_2$, the prefix *hypo* precedes the "chlorite"; as hence, *potassium hypochlorite*. To indicate the greater content of oxygen in $KClO_4$ (as compared to $KClO_3$), the prefix *per* precedes the "chlorate"; hence, *potassium perchlorate*.

It must be conceded, however, that such relationships are not always so easily discernible by inspection; thus

$$Na_2MnO_4 \text{ is sodium manganate}$$

and

$$NaMnO_4 \text{ is sodium permanganate.}$$

At a later point, we apply the concept of *oxidation state* which, in revealing the "apparent" charge of the central atom — in the cases just cited, manganese, Mn — proves to be the deciding factor in the choice of both prefix and suffix. Nomenclature is described as the occasion arises, particularly in view of its vagaries. Suffice it to say here that the nomenclature adopted to cover these situations conforms to the increases in the state of oxidation of the central atom. Consequently, for the various illustrations we have given, we imply that

$$\begin{bmatrix} \text{oxidation state} \\ \text{of Cl in} \\ \text{hypochlorite} \end{bmatrix} < \begin{bmatrix} \text{oxidation state} \\ \text{of Cl in} \\ \text{chlorite} \end{bmatrix} < \begin{bmatrix} \text{oxidation state} \\ \text{of Cl in} \\ \text{chlorate} \end{bmatrix}$$

$$< \begin{bmatrix} \text{oxidation state} \\ \text{of Cl in} \\ \text{perchlorate} \end{bmatrix}$$

and

$$\begin{bmatrix} \text{oxidation state} \\ \text{of Mn in} \\ \text{manganate} \end{bmatrix} < \begin{bmatrix} \text{oxidation state} \\ \text{of Mn in} \\ \text{permanganate} \end{bmatrix}.$$

In the same vein, occasional need to distinguish the variable valences of the metallic element in binary or ternary compounds invites further modifications. The suffixes *ous* and *ic* are attached to the root names of elements to indicate the variabilities of the valence. Illustratively,

$FeCl_2$ is ferr*ous* chloride (lower state of oxidation of iron, 2)
$FeCl_3$ is ferr*ic* chloride (higher state of oxidation of iron, 3)

and

$SnCl_2$ is stann*ous* chloride (lower state of oxidation of tin, 2)
$SnCl_4$ is stann*ic* chloride (higher state of oxidation of tin, 4).

Equivalent expressions for the iron and tin compounds just cited employ bracketed Roman numerals; thus

$FeCl_2$ may be designated as iron(II) chloride
$FeCl_3$ may be designated as iron(III) chloride

and

$SnCl_2$ may be designated as tin(II) chloride
$SnCl_4$ may be designated as tin(IV) chloride.

The familiar terminological alternations that apply to *acids* is comprehended in the following — as a last item in our immediate attention to nomenclature:

$HClO$, hydrogen hypochlorite, is correctly hypochlorous acid
$HClO_2$, hydrogen chlorite, is correctly chlorous acid
$HClO_3$, hydrogen chlorate, is correctly chloric acid
$HClO_4$, hydrogen perchlorate, is correctly perchloric acid

and

H_2MnO_4, hydrogen manganate, is correctly manganic acid
$HMnO_4$, hydrogen permanganate, is correctly permanganic acid.

Facility in the chemist's language, that is, in the writing of formulas, and fluency in reading and interpreting them becomes, as in any other language, virtually instinctive with habitual use. It is most desirable that the student commit to memory the more important valences commonly encountered, as interpreted from the electronic structures attributed to the atoms of interest. And certain refined differentiations that exist between "valence" and the nearly (but not quite) similar concept of "oxidation state" are likewise to be grasped.

TABLE 6:1

Common Valences of Some Elements in Binary Compounds

1	2	3	4
Metals:			
Lithium, Li	Magnesium, Mg	Bismuth, Bi	
Sodium, Na	Calcium, Ca	Aluminum, Al	
Potassium, K	Strontium, Sr	Chromium, Cr	
Rubidium, Rb	Barium, Ba		
Cesium, Cs	Cadmium, Cd		
Silver, Ag			
*Copper, Cu	*Copper, Cu		
*Mercury, Hg	*Mercury, Hg		
	*Cobalt, Co	*Cobalt, Co	
	*Iron, Fe	*Iron, Fe	
	*Manganese, Mn		*Manganese, Mn
	*Tin, Sn		*Tin, Sn
	Lead, Pb		
	Zinc, Zn		
Nonmetals:			
Fluorine, F	Oxygen, O	Arsenic, As	Carbon, C
Chlorine, Cl	Sulfur, S	(semi-metal)	Silicon, Si
Bromine, Br		Antimony, Sb	
Iodine, I		(semi-metal)	

*More important of variable valences of these elements in compounds.

TABLE 6:2

Common Valences of Some Polyatomic Ions (Radicals)

1	2	3
Cyanide, CN^-	Chromate, CrO_4^{2-}	Phosphate, PO_4^{3-}
Hydroxide, OH^-	Dichromate, $Cr_2O_7^{2-}$	Arsenate, AsO_4^{3-}
Nitrite, NO_2^-	Monohydrogen Phosphate, HPO_4^{2-}	
Nitrate, NO_3^-	Acetate, $C_2H_3O_2^-$	
Hydrogen Sulfate, HSO_4^-	Sulfite, SO_3^{2-}	
Hydrogen Carbonate, HCO_3^-	Sulfate, SO_4^{2-}	
Hyoochlorite, OCl^-	Silicate, SiO_3^{2-}	
Chlorate, ClO_3^-	Thioulfate, $S_2O_3^{2-}$	
Perchlorate, ClO_4^-	Carbonate, CO_3^{2-}	
Permanganate, MnO_4^-		
Dihydrogen Phosphate, $H_2PO_4^-$		
Ammonium, NH_4^+		

IMPLICATIONS OF EQUATIONS

The balanced chemical equation is a condensed statement that expresses, by symbols and formulas, both qualitative and quantitative information about a specific chemical reaction. It tells us, with direct simplicity, the specific relationships in weight and/or volume that exist between and among the initial reactants and the chemical products into which they are being converted. In our immediate concern with the balanced equation that affords these ultimate stoichiometric relationships, we necessarily seek a procedure that correctly establishes the coefficients of the equation; that is, the arithmetical multipliers of the different chemical units therein.

To be stoichiometrically valid, every chemical equation must fulfill all the following considerations:

1. The equation must express what actually happens *experimentally*. The chemical identities and chemical formulas of all species participating in change must be known and be written in their exact elemental, molecular, or ionic identities. If such identities have not been established by experiment, their empirical designations may be employed. In depicting the changes in identity that are being effected, it is to be noted that, as a rule, the equation is purely a formulation of net change; that is, it concerns itself solely with the species initially taken (appearing on the left of the arrow) and the final products as experimentally determined (appearing on the right). Frequently, little is known concerning the precise mechanism by which a particular substance undergoes alteration of its electronic make-up, but in nearly all cases it occurs virtually instantaneously. Prognostications of the actual step-by-step intermediate ("in-transit") changes in the chemical identities of species as they pass to the ultimate products are the difficult derivations sought by the chemical kineticist.

Essentially, then, the net balanced equation merely notes the disappearance or depletion of initial reactants and the formation or maintenance of final products. In the following illustration

$$A + B \rightarrow AB \text{ (intermediate)}$$
$$AB + B \rightarrow AB_2 \text{ (intermediate)}$$
$$\overline{A + 2B \rightarrow AB_2 \text{ (net)}}$$

The fact that species A has not combined with species B directly to form AB_2, but indirectly, through the intermediate formation of AB, has not altered the over-all net reaction of change that is being sought in the balanced chemical equation.

2. The equation must be balanced *atomically*. In conformity with the law of conservation of mass, equal numbers of the same kinds of all atoms must appear on both sides of the arrow in the equation.

3. The equation must be balanced *electronically*. In conformity with the law of conservation of energy, total net electrical charge, apparent or real, of the reacting materials specified on the left must equal exactly the total net electrical charge of the products specified on the right. As the balancing of changes in electrons — losses with gains — that result from breaking old and making new chemical bonds can be rather complicated, we appro-

priately develop systematic algebraic procedures that will ensure the satisfying of all conservation of energy.

METATHESIS AND REDOX

There are two broad categories within which all chemical reactions may be placed; *metathesis*, and *redox* (a contraction of the terms *reduction* and *oxidation*). The differences between them are based not upon the actual chemical identities of the substances involved but rather upon the algebraic involvements of the specific subatomic particles — electrons — that numerically determine the chemical characteristics of all identifiable matter.

Metathesis

In metathesis, as a net effect, no electrons are actually lost or gained by the substances undergoing mutual changes. Metathetically involved substances merely exchange chemical partners, and each of the partners retains the full complement of electrons and chemical bonds in its unit make-up that it had before the exchange. This may readily be visualized in the following metathetical exchange:

silver nitrate + sodium chloride → silver chloride + sodium nitrate

which is symbolized by

$$AgNO_3 \quad + \quad NaCl \quad \rightarrow \quad AgCl \quad + \quad NaNO_3.$$

Redox

In redox reactions, on the other hand, some substance(s) must lose electrons permanently — process called *oxidation* — and other(s) gain electrons permanently — process called *reduction*. Conservation of energy requires that the total number of electrons gained be exactly equal to the total number lost; that is, transference without dissipation. Moreover, because all matter has over-all (or net) electroneutrality, it is clear that the mechanisms of oxidation and reduction must occur simultaneously and that neither can occur without the other.

It is this double reaction that is being recognized in the contracted term "redox." The systematized algebraic "bookkeeping" of electrons that permits confident balancing of an equation for even the most complicated reaction is interpreted later. In order to facilitate progress to the primary objective of this chapter — the weight/volume relationships of the balanced equation — we confine ourselves for the moment to delineations of formulation that may readily be perceived by simple inspection. To be sure, of course, the simplest reactions are by no means necessarily metathetical.

Thus, although anyone could perceive immediately the correctness of the balancing in

$$C \ + \ O_2 \ \rightarrow \ CO_2$$

Carbon   Oxygen   Carbon dioxide

or require algebraic aids for application to

$$2H_2O \rightarrow \ 2H_2 \ + \ O_2$$

Water   Hydrogen   Oxygen

each, nonetheless, is somewhat subtly a redox reaction.

WEIGHT-VOLUME BY INSPECTION

We concern ourselves here with the reaction between copper sulfide (solid) and oxygen (gas) at high temperature to form copper oxide (solid) and sulfur dioxide (gas). Before making any stipulations with respect to its stoichiometry, we will derive the balanced equation for the reaction, in stepwise fashion:

1. Write all the reactants and products in accordance with their experimentally established chemical identities. This gives the provisional skeletal equation (yet to be balanced, that is); thus

$$CuS \ + \ O_2 \ \rightarrow \ CuO \ + \ SO_2$$

copper sulfide   oxygen   copper oxide   sulfur dioxide
solid   *gas*   *solid*   *gas*

2. Balance the atoms on either side of the arrow. We observe numerical equalities between the copper atoms at left and right; and similarly, between the sulfur atoms at left and right. But, the oxygen atoms must be balanced, there being 2 in total on the left and 3 in total on the right. Consequently, multiplying the 2 atoms on the left by a factor of $\frac{3}{2}$ will equalize the numbers of oxygen atoms at 3 on either side.

Therefore

$$CuS + \tfrac{3}{2}O_2 \rightarrow CuO + SO_2$$

This equation is now arithmetically balanced but it violates the convention that the coefficients (the multipliers that we place in front of the formula units) be *whole* numbers; hence,

3. Multiply the *entire* equation, left and right by 2 in order to convert the fraction $\frac{3}{2}$ to a whole number. We thus obtain

$$2CuS + 3O_2 \rightarrow 2CuO + 2SO_2$$

4. Observe whether any net electrical charge that might appear on species on one side of the equation is arithmetically compensated on the other.

In this case, no charges are involved, as would be expected of any compound or element that is written molecularly; for all molecules are electrically net-neutral. The equation fulfills, therefore, all requirements for conservation of energy.

The emphasis that has been placed upon this step 4 is not without ample justification. Note, for example, the pitfalls that await the unwary who would be content merely with equal numbers of atoms on both sides of the following equation without concern for the requirement to equalize electrons. Thus

$$Al^0 \; + Cu^{2+} \rightarrow \; Al^{3+} \; + \; Cu^0$$

Aluminum   Copper   Aluminum   Copper
metal   ion   ion   metal

As is, this is a most inaccurate equation, for there is a 2^+ charge on the left and a 3^+ charge on the right. Clearly, to reconcile the discrepancies in balance of energy, the Cu^{2+} must be multiplied by 3 and the Al^{3+} by 2; and this step concomitantly changes the mass balance of atoms. To restore mass balance, the equation should read

$$2Al^0 + 3Cu^{2+} \rightarrow 2Al^{3+} + 3Cu^0,$$

in order to reveal, correctly, the precise atomic and electronic stoichiometry of the reaction.

Returning now to our original illustration, the relationships to be perceived between and among the weights and/or volumes of reactants and products in the reaction may be enumerated as shown in Table 6:3:

TABLE 6:3

Relationships between Reactants and Products

Reaction equation:	2CuS (solid)	+ 3O$_2$ (gas)	⟶ 2CuO (solid)	+ 2SO$_2$ (gas)
Formula units:	$(2 \times 6.02 \times 10^{23})$:	$(3 \times 6.02 \times 10^{23})$:	$(2 \times 6.02 \times 10^{23})$:	$(2 \times 6.02 \times 10^{23})$
Grams:	$\begin{bmatrix} 2(63.54 + \\ 32.064) \\ = 191.21 \end{bmatrix}$	$\begin{bmatrix} 3(2 \times 15.9994) \\ = 95.9964 \end{bmatrix}$	$\begin{bmatrix} 2(63.54 + \\ 15.9994) \\ = 159.08 \end{bmatrix}$	$\begin{bmatrix} 2(32.064) + \\ 2(2 \times 15.9994) \\ = 128.126 \end{bmatrix}$
Liters at STP:	— (not applicable to solids)	$\begin{bmatrix} 3 \times 22.4 \\ = 67.2 \end{bmatrix}$	— (not applicable to solids)	$\begin{bmatrix} 2 \times 22.4 \\ = 44.8 \end{bmatrix}$

The practical applications of this reaction are interpreted in the problems stated and solved in the following section.

SAMPLE CALCULATIONS

(All reactions proceed in accordance with the one
just illustrated in the preceding section.)

■ PROBLEM 1: To calculate the number of grams of oxygen gas required to produce 30.0 g of copper oxide.

Solution. The equation for the reaction shows that

159.08 g of CuO are produced from 95.9964 g of O_2

Hence,

$$1 \text{ g of CuO is produced from } \frac{95.9964}{159.08} \text{ g of } O_2$$

and, consequently

$$30.0 \text{ g of CuO are produced from } \left(30 \times \frac{95.9964}{159.08}\right) \text{ g of } O_2$$

$$= 18.1 \text{ g of } O_2. \text{ (ANSWER)}$$

■ PROBLEM 2: To calculate the number of liters of sulfur dioxide gas produced at STP from 40.0 g of copper sulfide.

Solution. The equation for the reaction shows that at STP

191.21 g of CuS produce 44.8 liters of SO_2

Hence,

$$1 \text{ g of CuS produces } \frac{44.8}{191.21} \text{ liters of } SO_2$$

and, consequently

$$40.0 \text{ g of CuS produce } \left(40.0 \times \frac{44.8}{191.21}\right) \text{ liters of } SO_2$$

$$= 9.37 \text{ liters } SO_2 \text{ at STP. (ANSWER)}$$

■ PROBLEM 3: To calculate the number of liters of oxygen gas consumed, and of sulfur dioxide gas produced, both at STP, when 50.0 g of copper sulfide have been converted to copper oxide.

Solution. The equation for the reaction informs us that at STP

191.21 g of CuS consumes 67.2 liters of O_2 and

produces 44.8 liters of SO_2.

Hence,

1 g of CuS consumes $\dfrac{67.2}{191.21}$ liters of O_2 and

$$\text{produces } \dfrac{44.8}{191.21} \text{ liters of } SO_2.$$

Consequently,

50.0 g of CuS consume $\left(50.0 \times \dfrac{67.2}{191.21}\right)$ liters O_2 and

$$\text{produce } \left(50.0 \times \dfrac{44.8}{191.21}\right) \text{ liters } SO_2.$$

The equated volumes of gases

$$= 17.6 \text{ liters } O_2 \text{ at STP}$$
$$= 11.7 \text{ liters } SO_2 \text{ at STP.} \quad \text{(ANSWERS)}$$

Although we have done so as a concession to introductory clarification, it really was not necessary to work out both volumes in this precisely similar fashion. Had we ascertained the volume of only one by this procedure, the other would automatically have checked out in the ratio of the volumes expressed (under identical conditions of temperature and pressure) by the coefficients of the gases in the balanced equation. Thus, the volume of oxygen gas consumed would have to be $\frac{3}{2}$ of the volume of the sulfur dioxide gas produced; or, alternatively, the volume of the sulfur dioxide gas produced would have to be $\frac{2}{3}$ of the volume of oxygen gas consumed. This relationship checks correctly with the volumes that have been calculated for each; for

$$O_2 \text{ consumed at STP} = \tfrac{3}{2} \times 11.7 \text{ liters of } SO_2 \text{ produced at STP}$$
$$= 17.6 \text{ liters, as before}$$

and

$$SO_2 \text{ produced at STP} = \tfrac{2}{3} \times 17.6 \text{ liters of } O_2 \text{ consumed at STP}$$
$$= 11.7 \text{ liters, as before.}$$

Nonstandard states of temperature and pressure offer no complications to stoichiometric weight-volume or volume-volume relationships. The variable conversions of volumes to and from STP that may be required by the specific problem are readily made by the mathematical applications of Boyle's and Charles' laws. This is a highly familiar practice in the initiation of basic chemistry.

CHEMICAL EQUIVALENCE

All stoichiometric calculations, whatever the nature of the reaction may always be made directly from the balanced equation, as demonstrated. Yet many reactions, notably those of redox, readily lend themselves computatively to the far more convenient stoichiometric process called *chemical equivalence*. This algebraic concept, which bypasses completely the balancing of equations, is advantageously applied in a later chapter dealing with volumetric stoichiometry. Our purpose here in opening the door a bit just for a quick "peek" is merely to acquaint the student with the relationship of this concept to the implications of the balancing process that yields the stoichiometric equation.

To be sure, although utilizing the *equivalent* obviates the need for balancing an equation, it is still indispensable to know precisely, as in the equation itself, the chemical identities, in formulas, of the substances of immediate interest that undergo change, and of the products into which they are converted. The point of convenience, however, is that the concept of equivalence applies itself, stoichiometrically, solely to the atom(s) that undergo changes in valence or state of oxidation. Inasmuch as nearly all balanced equations depict one or more atoms in interacting substances whose valences or oxidation states have not changed in the chemical transformations of one species to another, the advantage of thus fixing attention on those that have is frequently an extremely worthwhile laborsaving device.

The objective of chemical equivalence is to provide common units that will express the different weights of different substances that react with each other, and also denote a 1 : 1 inter-reaction ratio between such units; or any multiple of this ratio, fractional or integral, that similarly expresses equality.

Let us interpret chemical equivalence by first inspecting the following equations:

$$\underset{\text{Magnesium}}{Mg} \; + \; \underset{\substack{\text{Hydrochloric}\\\text{acid}}}{2HCl} \; \rightarrow \; \underset{\substack{\text{Magnesium}\\\text{chloride}}}{MgCl_2} \; + \; \underset{\text{Hydrogen}}{H_2}$$

$$\underset{\text{Magnesium}}{Mg} \; + \; \underset{\substack{\text{Sulfuric}\\\text{acid}}}{H_2SO_4} \; \rightarrow \; \underset{\substack{\text{Magnesium}\\\text{sulfate}}}{MgSO_4} \; + \; \underset{\text{Hydrogen}}{H_2}$$

$$\underset{\text{Magnesium}}{2Mg} \; + \; \underset{\substack{\text{Phosphoric}\\\text{acid}}}{2H_3PO_4} \; \rightarrow \; \underset{\substack{\text{Magnesium}\\\text{phosphate}}}{Mg_3(PO_4)_2} \; + \; \underset{\text{Hydrogen}}{3H_2}$$

Let us focus attention upon the common reactant, magnesium. In each instance, it is observed that one mole of the metal reacts with a different number of moles of acid, the respective ratios being:

$$1 \text{ mole Mg} : 2 \text{ moles HCl}$$
$$1 \text{ mole Mg} : 1 \text{ mole } H_2SO_4$$
$$1 \text{ mole Mg} : \tfrac{2}{3} \text{ mole } H_3PO_4.$$

It can never be correctly contended that reaction necessarily takes place mole for mole. Yet there is a common resemblance between the two reactants in each of the three different cases; namely, the numerical equality of the total numbers of valence bonds that the reactants reveal (inclusive of the reaction coefficient preceding each) with respect to their capacity either to combine with, or supply, or otherwise react with hydrogen atoms. This has already been discussed as the *standard of valence* — unity — which, as likewise already defined, is replaceable by any other atom(s) of unit valence in actual combination or supply. As the chlorine atom (first equation) reveals a valence of 1 in thus combining with one atom of hydrogen (from HCl), the formation of $MgCl_2$ establishes a valence of 2 for the reacting Mg. Note, then, the numerical identity between the reactants: two valence bonds are also provided by the reacting 2HCl.

If we now also set the standard of chemical equivalence as that of a single hydrogen atom (at unity), then, clearly, the gram-equivalent weight of any substance must be that part of its formula weight which combines with, supplies, or otherwise reacts with 1 gram-atomic weight of hydrogen; or with 1 g-atomic weight of any other atom that, itself chemically combines with, supplies, or otherwise reacts with one hydrogen atom. Therefore, as Mg here contains 2 equivalents, it follows that

$$1 \text{ g-equiv weight Mg} = \frac{\text{g-formula weight Mg}}{2}$$

$$= \frac{24.312}{2}$$

$$= 12.156 \text{ g}$$

and, as HCl contains 1 equivalent, it follows that

$$1 \text{ g-equiv weight HCl} = \frac{\text{g-formula weight HCl}}{1}$$

$$= \frac{36.461}{1}$$

$$= 36.461 \text{ g.}$$

That is, because all chemical reaction occurs *equivalent for equivalent*, 12.156 g of Mg (1 g-equiv) reacts exactly with 36.461 g of HCl (1 g-equiv), or in any corresponding ratio, to yield the products stated.

In similar fashion, there being two equivalents each of Mg and of H_2SO_4 (second equation), stoichiometry of gram-equivalent weights must provide that as

$$1 \text{ g-equiv weight Mg} = \frac{\text{g-formula weight Mg}}{2}$$

$$= \frac{24.312}{2} = 12.156 \text{ g}$$

and

$$1 \text{ g-equiv weight } H_2SO_4 = \frac{\text{g-formula weight } H_2SO_4}{2}$$

$$= \frac{98.978}{2} = 49.039 \text{ g},$$

12.156 g Mg (1 g-equiv) reacts exactly with 49.039 g H_2SO_4 (1 g-equiv), or in any corresponding ratio, to yield the stated products.

The third reaction shows Mg still with its chemical equivalence of 2, and H_3PO_4 with a chemical equivalence of 3. Consequently,

$$1 \text{ g-equiv weight Mg} = \frac{\text{g-formula weight Mg}}{2}$$

$$= \frac{24.312}{2} = 12.156 \text{ g}$$

$$1 \text{ g-equiv weight } H_3PO_4 = \frac{\text{g-formula weight } H_3PO_4}{3}$$

$$= \frac{97.9953}{3} = 32.6651 \text{ g},$$

and 12.156 g of Mg (1 g-equiv) reacts exactly with 32.6651 g of H_3PO_4 (1 g-equiv), or in any corresponding ratio, to yield the stated products.

The more subtle involvements of variable equivalent weights of the same substance, which reflects its different degrees of bonding (multiple valence), receives appropriate attention in the chapter dealing with the balancing of the redox equation. Briefly, here, delineations to be made of the necessary algebraic methods of balancing a redox equation will make it readily evident that it is well to avoid the writing of equations for highly complex reactions when the sole concern is with their relationships of weight and/or volume.

The character of the equations so far offered — simplicity consonant with the need for introductory clarity — should not invite the idea that "it is just as fast to work it out from the equation." And particularly so when all products necessary for the equation are not known or not provided. Nonetheless, the writing of correctly balanced equations is a "must" for the literate chemist. He would perceive in them not only an exact description of the stoichiometry of a reaction, but also the means for appraising the thermochemical and photochemical energy involved, and plausibly elucidating the rate and possible mechanism of the reaction.

CHAPTER 7

BALANCING THE REDOX EQUATION

The predominant objective of this chapter is the discerning application of the two important mathematically organized methods in general use for balancing the oxidation-reduction equation:

1. by *oxidation number;*
2. by *ion-electron* method (also called "half-reactions" and "partials").

Intensive interpretations of both afford not only prudent appraisals of their relative merits and preferential applications in specific situations, but also optimum assurances that even the most complex of reaction equations may be balanced with facility and correctness. Indeed, the frequent frustrations of time-extravagant and unrewarding haphazard "trial and error" methods amply justify all efforts expended in this present pursuit.

ELECTRON-TRANSFER IN OXIDATION AND REDUCTION

Careful distinction must be made between *oxidation*, the process, and *oxidant*, the oxidizing agent. *Oxidation* is the net over-all process by which a species entering into chemical change loses one or more of the electrons that it had initially; or, equally valid as an observation, the process whereby its *state of oxidation* is increased. The *oxidant*, on the other hand, being the particular reacting species that induces the loss of the electrons (which it thereupon acquires) is consequently reduced in the process. The oxidation state of the oxidant must therefore be diminished. Conversely, *reduction*, as a process, signifies the net gain of electrons by a reacting species, with resultant decrease in its oxidation state. The *reductant* (reducing agent), however, in functioning as the particular species that actually gives up or transfers from itself one or more of its electrons, must become oxidized in the process; hence, its oxidation state must be increased.

Because, in oxidation-reduction, numbers of electrons lost and gained must be equal (transference without dissipation), the total diminution in the

128

oxidation states of all of the atoms of the oxidant that enter into reaction (which are reduced) must equal exactly the total increase in the oxidation states of all of the atoms of the reductant that are simultaneously oxidized. The oxidant thus acquires the exact number of electrons released by the reductant.

It is precisely this awareness of the quantitative character of electron transference (experimentally delineated and validated in Faraday's laws of electrolysis) that provides the highly useful means for balancing even the most complex and involved equations of reaction. The device here is simply to keep track of, and accurately account for, all electrical charge or changes in the oxidation states of the atoms, and to make the necessary equalizing adjustments represented by the coefficients of the chemical species in the written equation.

Normally the oxidant and reductant involved in reaction are of different chemical identities, but it nevertheless is possible for the same substance to act simultaneously as both oxidant and reductant in a specific reaction. This type of redox, in which a substance simultaneously oxidizes and reduces itself, is called *disproportionation* (also known by the terms "internal-," "autonomous-" or "self-oxidation-reduction").

OXIDATION STATE VERSUS VALENCE

The terms "oxidation state" and "valence" are frequently confused or quite loosely used. Even in the awareness of the existence of a very real and valid distinction, "oxidation state" expresses the apparent, frequently arbitrary, charge that an atom of an element in a particular species is deemed to contribute to the over-all net charge of the chemical formula unit of which it is a constituent part. Although the same element may have many different oxidation states in various substances, the magnitude of its charge in each instance, is always expressible by a number (not necessarily an integral whole number). It also has the associated algebraic sign, either plus $(+)$ or minus $(-)$ that denotes its electrical character in chemical bonding, in conformity with the concept that it either gains or loses electrons in the convenient "counting process."

Valence, or combining capacity, on the other hand, is always a pure integral whole number. It is never fractional and never represented with either a negative or a positive charge, inasmuch as it expresses solely the number of bonds in which a given species (atom, simple or polyatomic ion, molecule, or empirical formula) is visualized to participate during electrovalent or normal or coordinate covalent bondings.

Oxidation state and valence may differ from each other even in numerical magnitude; for example, the oxygen atom in hydrogen peroxide, H_2O_2. Here, the electronic structure

$$\left[H \; \overset{\times\times}{\underset{\times\times}{O}} \overset{\times\times}{\underset{\times\times}{O}} \; H \right] \text{ for } H_2O_2$$

clearly reveals each oxygen atom bonded in two positions; hence, each *peroxy* oxygen atom has a valence of 2. Yet, the oxidation state of the peroxy oxygen atom is -1, as is presently mathematically developed. In the compounds called "superoxides," such as KO_2 (not to be confused with the familiar K_2O), the oxidation state of the oxygen atom is $-\frac{1}{2}$. As another illustration, with carbon atoms: the oxidation state of the carbon atom in CH_2O is zero; in C_2H_2, each carbon atom is in the oxidation state of -1; in C_2H_4 each carbon atom is in -2 oxidation state; in C_2H_6 the carbon atom has achieved a -3 oxidation state; and in CH_4 the oxidation state of the carbon is -4. Nonetheless, the valence of the carbon atom remains 4 in all of these compounds.

Though assignments of oxidation state may be artificial, they serve profitably in the "bookkeeping" procedures utilized for balancing the equations of reactions, between nonelectrolyte species in particular, as well as for all reactions that occur in the molten (fused) states of materials. Its general application to electrolytes is likewise equally valid, although the equations of reactions between electrolytes in aqueous media are balanced more rewardingly (in the actual chemistry of behavior of the species) by an alternate ion-electron or half-reaction method, which avoids the artificial assignments of the oxidation state. This is explained later.

The utility of valence in chemical bonding is perhaps best served by reserving the term for the delineations of the actual over-all net charge upon a simple or complex ionic species, and confining the assignments of oxidation state to the constituent atoms that individually comprise the whole. Thus, the dichromate ion, $Cr_2O_7^{2-}$ may accurately be said to have a valence of 2 (an oxidation state of -2, if desired), and the chromium and oxygen atoms then are in characteristic oxidation states that yield, by collective algebraic addition, the net charge (valence) of the dichromate ion as depicted.

OXIDATION STATE AND OXIDATION NUMBER

A careful distinction must be made between oxidation *state* and oxidation *number*. *Oxidation state* refers to the charge, apparent or real, upon a single atom. Upon occasion, an average charge is assigned to the single atom when two or more atoms of a given element are present in different oxidation states in the same chemical formula unit.

Oxidation number connotes the cumulative total of the oxidation states of all the atoms of an *identical* element present in a formula unit. Consequently, the numerical values of oxidation state and oxidation number are very frequently quite different from each other. Oxidation is precisely equal to oxidation number when just one atom of an element is present in the formula.

A few illustrations make these distinctions clear. In potassium chromate, K_2CrO_4, the oxidation states of the elements are as follows: $K = +1$, $Cr = +6$, $O = -2$; the oxidation numbers, however are $K = +2$, $Cr = +6$, $O = -8$. In the dichromate ion, $Cr_2O_7^{2-}$, the oxidation *states* are $Cr = +6$, $O = -2$; the oxidation numbers for a formula unit must therefore be $Cr = +12$, $O = -14$.

Witness that the algebraic addition of these oxidation numbers yields the net negative charge of 2 upon the complete ion itself, which is its valence as an anion.

In the simple chromic ion, Cr^{3+}, oxidation state and oxidation number are numerically identical; that is, $+3$ which is equal to the positive charge upon the simple cation itself, and its valence is also 3.

The significance of oxidation number is thus revealed by the fact that the species conforms in each case to the following rule:

The algebraic sum of all of the oxidation numbers of the atoms of all elements in a molecule (an electrically uncharged formula unit) must be equal to zero; and, the algebraic sum of all of the oxidation numbers of the atoms of all elements constituently present in a net electrically charged ionic species must be equal to the net charge upon the species; that is, equal numerically to the valence of the entire species.

The utility of oxidation number for balancing equations is further enhanced by its special applicability to calculations of gram-equivalent weights, both for metathetical and for redox stoichiometry, without the need for previously balancing a complete and comprehensive equation for the reaction under consideration. This has already been given attention.

It becomes apparent that when oxidation-reduction occurs the net physical process and total over-all physical mechanism, regardless of the step-wise chemical intermediates (frequently several, particularly in relatively slow or noninstantaneous reactions), must logically and inevitably involve a complete transference of electrons from the reducing species to the oxidizing species. This is amply demonstrated in the galvanic (voltaic) cells, wherein the same net chemical transformations that occur in species intimately mixed in conventional experiments with test tube and beaker are duplicated, even though the elements are now separated in space — in separate compartments, by salt bridge, inert electrodes, connecting wire for external circuit. That the comparable products formed in the separate compartments of such cells and in the intimate mixtures formed in test tubes or beakers are identical is convincingly demonstrated by appropriate chemical tests. The actual transference of the electrons is likewise realistically observed and validly interpreted in the deflection of the needle of the galvanometer or voltmeter placed in the external electrical circuit of the cell.

DERIVATIONS OF OXIDATION NUMBERS

Oxidation numbers are fundamentally rooted in the concepts of the relative electronegativity of atoms, and of the octet theory of structure and stability of species. In essence, to each and every atom, whether present alone or as part of a polyatomic species (ionic or molecular), is assigned the electrical charge (its oxidation state) presumed to represent its net electrical contribution to the over-all charge of the chemical formula unit as a whole. The means used to fulfill objectives here is to count with the more electronegative atom all of the electrons that this atom shares with others in chemical bondings. This means that, the more electropositive atom is solely for purposes of counting, shorn of the electron(s) it brought with it into the "partnership."

Thus, in the conventional portrayal of a neutral atom, the symbol of the element itself represents the kernel of the atom; that is, its nucleus plus all its electrons except those on the outermost energy-level — the valence electrons. By conveniently deploying dots, circles, or crosses to indicate the valence electrons, we may picture the "octet" structural configuration of the hydrogen chloride molecule as

$$\left[\text{H} \overset{\times\times}{\underset{\times\times}{\times}} \text{Cl} \overset{\times\times}{\underset{\times\times}{}} \right]$$

From this we deduce an oxidation state of -1 for the chlorine atom, and of $+1$ for the hydrogen atom, because the two electrons that are being shared must now be counted fully with the more electronegative chlorine atom. As both atoms, before their mutual bonding, were electrically neutral (oxidation state, zero) the counting process has left the hydrogen atom in the molecule of hydrogen chloride shorn of one electron because of its polar displacement toward the far more electronegative chlorine atom in the molecule. This brings about the incremental imbalance of a single unneutralized positive nuclear charge that is responsible for the oxidation state of the hydrogen atom, $+1$. The chlorine atom, now enriched with one incrementally-counted electron that is now not electrically neutralized by its own positive nuclear charge, is thus endowed with the single negative charge in excess that constitutes its oxidation state, -1.

In the formulations

$$\left[\overset{\times\times}{\underset{\times\times}{\times}} \text{Cl} \overset{\times}{} \quad \overset{\cdot\cdot}{\underset{\cdot\cdot}{}} \text{Cl} : \right]$$

and

$$[H^\times H]$$

as neither atom of the Cl_2 and of the H_2, respectively, can reasonably be assumed to be either more or less negative than the other, it is apparent that there is no shifting of electrons for counting purposes; hence each atom remains in an oxidation state of zero. This situation must prevail with respect to the atoms of all pure elements, regardless of the number of the similar atoms present. Thus all the individual sulfur atoms in S_2, S_4, S_6, and S_8 are in the oxidation state of zero, which is characteristic of monoatomic sulfur,

In H_2O, the structural formulation

$$\left[\begin{array}{c} {}^\times_\times O^\times\ H \\ {}^\times \\ H \end{array}\right]$$

clearly shows that the oxygen atom must have an oxidation state of -2, because it acquires for counting purposes one electron from each of two hydrogen atoms, each of which is now in an oxidation state of $+1$. In hydrogen peroxide, however, depicted structurally by the formulation

$$\left[H\ {}^\times_\times O^\times_\times O^\times_\times\ H \right]$$

although the sharing of electrons between the two oxygen atoms does not alter their initial oxidation states of zero, note that each oxygen atom gains one electron for counting purposes from the hydrogen atom with which it is bonded. Hence, the peroxy-oxygen atom is in an oxidation state of -1.

The identical determination would have been made had the oxygen atoms been evaluated directly from the free peroxide ion, O_2^{2-}, as available from ionic Na_2O_2, sodium peroxide. In the electronic formulation of this compound,

$$2Na^+\left[{}^\times_\times O^\times_\times O^\times_\times \right]^{2-}$$

the net charge of 2^- for the peroxide ion as a whole, as acquired by the electrovalent transfer of one electron from each of the two sodium atoms, must be preserved. Each of the two bonded and equally electronegative oxygen atoms must, therefore, be regarded as contributing equally to this collective negative charge; consequently, each is in the oxidation state of -1.

It would be presumed that bondings between atoms of different electronegatives always give rise to oxidation states greater than or less than the zero value which is assigned to uncombined atoms or to those that are bonded with others of similar identity and electronegativity. Quite deceptively, on occasions, no change need be conceived despite the known presence of polar bondings. This inconsistency arises unmistakably from the artificiality of the octet theory as a concept of chemical principle — even though it is profitably useful as an expediently simple device in largely reconciling the structural bondings of species with their chemical behavior. The associated limitations of atomic orbitals that encourage the visualizing of the electrons being contributed by each atom to the over-all identity of a particular chemical species, as still belonging to such atom, rather than to the entire species as a collective whole unit, have hardly proved conducive to realistic and truly interpretative portrayals. These are within the province of the far more complicated concepts of the molecular orbital theory.

Illustrative of this quite arbitrary and oversimplified situation, resulting from the atomic orbital concept, is the compound formaldehyde, CH_2O. Assigning the oxidation states of -2 to oxygen and $+1$ to hydrogen here must leave the element carbon, despite its polar bonding, with an oxidation state of zero — the same assignment as it would be given as a neutral, uncombined atom. Thus, quite unrealistically, it is assumed that the carbon atom in this compound has contributed nothing to the over-all electronic makeup of the compound.

In a similar vein, realism in the concept of atomic orbitals is hardly further enhanced by the presence of fractional oxidation states. Thus, in the tetrathionate ion, $S_4O_6^{2-}$, the assignment to oxygen of the oxidation state of -2 and the oxidation number of -12 inevitably leaves the sulfur with an oxidation number of $+10$, because the net charge on the entire ion itself — its valence — must be conserved. It suggests, moreover, that the sulfur atoms are not *all* in the same oxidation state, but rather in a $+\frac{5}{2}$ representative of the average. Structural evidence is available that substantiates this because three of the four sulfur atoms actually are in different bonding positions with respect to the more electronegative oxygen atoms.

In reverse, however, an integral whole number assigned as an oxidation state does not necessarily mean that similar chemical atoms in a particular compound are in the same oxidation state. Thus the nitrogen atom in the molecule of nitrous oxide, N_2O, in conformity with the assigned rules of derivation, is in an oxidation state of $+1$, although an appraisal of the two "resonance" forms

$$\left[\overset{\times\times}{\underset{\times\times}{N}} N_{\times}^{\times} \ddot{O} : \right]$$

and

$$\left[{}_{\times}^{\times}N {}_{\times\times}^{\times\times} N {}_{\times}^{\times} \ddot{O} : \right]$$

clearly reveals the nonequivalent bonding positions of the two nitrogen atoms with respect to the oxygen atom. In these oversimplified visualizations of the compound, the nitrogen atom that is bonded to the oxygen may be regarded as having been deprived to two of its electrons (for counting purposes) by their polar shifting to the favored proximity of the more highly electronegative oxygen atom. This leaves the oxygen-bonded nitrogen atom in an oxidation state of $+2$. The nitrogen atom that is not oxygen-linked is conveniently presumed to share its electrons fairly equally in the nitrogen-nitrogen bond. This is rather unrealistic because, despite the identity of chemical relationship, obvious stresses are placed on its shared electrons by the oxygen-linked nitrogen atom, which has a partial increment of unrequited polar positive charge as a result of the displacement of its own electrons toward the oxygen atom. In essence, though, the initial oxidation state of zero of the nitrogen atom at the end may be regarded as unaltered; hence, the average of $+1$ that is obtained with the separate nitrogen atoms in oxidation states of 0 and $+2$, respectively.

CHANGES IN OXIDATION STATES — REDOXING

The balancing of equations is concerned critically with changes in oxidation states and in oxidation numbers. In conformity with the concepts already developed, it follows that if the thiosulfate ion, $S_2O_3^{2-}$, is converted to the tetrathionate ion, $S_4O_6^{2-}$, under suitable reaction conditions, there is a net gain of $+\frac{1}{2}$ in the oxidation state of a sulfur atom in the $S_2O_3^{2-}$ *ion* (from an initial $+2$ *to* $+\frac{5}{2}$ in the product); and, consequently, an increase of $+1$ ($+\frac{1}{2} \times 2$) in the oxidation number of this element in a formula unit. Were the conditions of reaction favorable to the conversion of the thiosulfate ion to sulfite ion, SO_3^{2-}, the sulfur in $S_2O_3^{2-}$ would gain $+2$ in its oxidation state (from its initial $+2$ to $+4$ in the product); and this time the oxidation number would increase $+4$ in the formula unit ($+2 \times 2$).

Confusion sometimes arises with validly different expressions of precisely the same thing. Thus, in the conversion of phosphine, PH_3, to orthophosphoric acid, H_3PO_4, the phosphorus atom in the former obviously gains $+8$ in oxidation state (from its initial -3 to $+5$ in the product). The same gain in oxidation state represented by this oxidation process of phosphorus in phosphine may also be depicted as a *loss* of -8 (*minus* 8) in oxidation state; because 8 units of negative charge (electrons) have been removed from the species. In neither statement is there any conflict with the concept of increase in absolute magnitude of the assigned value, because the assigned value means number plus charge (either positive or negative), considered as a collective unit in regard to oxidation and the consequent loss of electrons. Thus, $-3 + (+8)$ and $-3 - (-8)$ are the same in reconciling the *change* of the phosphorus atom from its -3 oxidation state in phosphine to its $+5$ oxidation state in H_3PO_4.

What has been stressed in the foregoing is what is inherent in the mutual compatibility of defining oxidation as being both an increase in absolute value of oxidation state and a decrease in the number of electrons retained by a species, each being a concomitant consequence of the other. By the same terms of definition, chemical reduction must be both a decrease in absolute value of oxidation state and an increase in the number of electrons held by a species after reaction — each, again, being a simultaneous consequence of the other.

In calculations of change, there must be no ambiguity in expressing what is being increased or decreased (lost or gained) — the oxidation state, or the oxidation number, or the number of electrons. Clarity may well be served in delineating required alterations in charge of species, by referring to such alterations as merely *units of change*.

The many obvious inconsistencies of the total picture notwithstanding, the method of balancing equations by oxidation number remains a completely successful mathematical device with reactions of all types both molecular and ionic, in liquid, in gaseous, and solid media. As an accounting procedure, it is a completely feasible and effective means for keeping track of changes in oxidation states and of ensuring that "books balance" on total oxidation numbers; that the *net total decrease precisely equals the net total increase.*

RULES FOR ASSIGNING OXIDATION STATE

As a very necessary preliminary to the actual formal balancing of equations by the method of oxidation numbers, the following rules of assigning oxidation state, although it is by no means comprehensive enough to meet all required situations, will prove exceedingly valuable in practically eliminating any necessity for the highly laborious derivations from structural electronic formulas that might otherwise be required:

1. *Atoms in free elements:* Oxidation state is always zero, whether the element be monatomic by empirical representation, or monatomic or polyatomic by actual molecular constitution.

2. *Atoms in simple, monatomic ions:* For Fe^{3+}, Cu^{2+}, Ag^+, Cl^-, S^{2-}, etc., oxidation state is always the charge upon the ion itself; the numerical magnitude of oxidation state is precisely equal to that of the valence.

3. *Atoms in polyatomic ions involving only one element:* For Hg_2^{2+}, O_2^{2-}, O_2^-, N_3^-, etc., the oxidation state of each atom is represented by its prorata share of the actual total charge upon the particular ion; thus, respectively, $+1$ for the mercury atom in the mercurous ion; -1 for the oxygen atom in the peroxide ion; $-\frac{1}{2}$ for the oxygen atom in the superoxide ion; and $-\frac{1}{3}$ for the nitrogen atom in the azide ion.

4. *Atoms in polyatomic species, ionic or molecular, involving more than one element*: the following simple secondary rules serve not only to establish directly the greater number of required oxidation states of atoms most frequently encountered in formulations, but also permit the quick derivation therefrom of many others that are needed. *These rules should be carefully committed to memory.*

(a) *The oxygen atom:* In all bondings with dissimilar atoms other than fluorine, oxidation state is -2. As the only element more electronegative than oxygen is fluorine, in any bonding with the latter the oxygen atom is left in a positive oxidation state. Thus, in the compound OF_2 an oxidation state of $+2$ must be assigned to the oxygen atom.

(b) *Periodic Group I atoms:* These comprise Li, Na, K, Rb, Cs, Fr. In all bondings with dissimilar atoms (bondings are necessarily electrovalent because of the low ionization potentials of these alkali metals) the oxidation state of each is $+1$.

(c) *Periodic Group II atoms:* These comprise Be, Mg, Ca, Sr, Ba, Ra. In all bondings with dissimilar atoms (bondings are primarily electrovalent, although an appreciable covalent character is noted for some beryllium and magnesium compounds) the oxidation state of each is $+2$.

(d) *The hydrogen atom:* In all bondings with dissimilar atoms, the oxidation state is $+1$, except in *hydrides* wherein the hydrogen atom is a partner in ionic bonding with a more electro-positive atom, generally an element of Periodic main groups I and II. In the hydride, consequently, the hydrogen atom, as the more electronegative individual, must be assigned the -1 oxidation state.

The wide diversity of oxidation states of most of the elements, which bespeaks the availability for bonding of multiple s and p electrons in their outer valence shells, and, in the transition elements, of d electrons as well. Any further effort in the direction of organized classification is unnecessarily involved, and even unwarranted, because in nearly all reactions of electrolytes they readily respond to simple derivation by utilizing charges of atoms that are already known.

Thus, in the formulas Na_2"X"O_4 or "X"O_4^{2-}, the oxidation state of the unknown "X" is $+6$ automatically, without knowledge of the chemical identity of "X." In $KHSO_4$, in conformity with the rules presented, the precise oxidation state of the sulfur atom here is, inescapably, $+6$, even though in the many different compounds that this element can form, the oxidation states of sulfur may range downwards as far as -2.

It also proves helpful to remember the frequently encountered ion-groupings that encompass various elements that are highly versatile in oxidation state, and which invariably act as tightly bonded groups (radicals). Representative of these are the chlorate ion, ClO_3^-; perchlorate ion, ClO_4^-; sulfite ion, SO_3^{2-}; sulfate ion, SO_4^{2-}; nitrate ion, NO_3^-; nitrite ion, NO_2^-; orthophosphate ion, PO_4^{3-}. The net charges of these collective units are always immediately deducible by inspection from the oxidation numbers assigned to hydrogen in the respective families of the acids: $H(ClO_3)$, $H(ClO_4)$, $H_2(SO_3)$, $H_2(SO_4)$, $H(NO_3)$, $H(NO_2)$, and $H_3(PO_4)$, respectively.

The ammonium ion "radical," NH_4^+, likewise are frequently encountered. A knowledge of the common group-association charges upon such radicals frequently enables the student to evaluate the oxidation state of a third atom concomitantly present, without the perplexity of deciding which of the variable oxidation states of chlorine, sulfur, nitrogen, phosphorus, or of other atoms of variable oxidation state in radicals is to be taken.

Thus, in $FeSO_4$, the oxidation state of the iron is $+2$; in $Fe_2(SO_4)_3$, the oxidation state of the iron is $+3$. Where the complexity of a species and the absence of familiar groups do not permit this selective determination of the specific assignment for an atom of variable oxidation state, it must be derived from a collective appraisal of configurations of atomic electrons, the relative positions of the bonding atoms in the periodic table, and experimental data covering chemical stabilities of the various oxidation states under the given conditions of the reaction.

SKELETAL EQUATION IN BALANCING OXIDATION NUMBERS

In the formal mechanism of balancing a redox equation by the method of oxidation numbers, the following are the procedural steps in their precise sequence of application. They should be carefully followed.

1. Ascertain by inspection and assign the values on both sides of the skeletal (unbalanced) equation for the oxidation state of all elements in species that are involved in mutual change of oxidation state. Remember that oxidation state is a determination for a single atom.

2. Determine, by comparing the respective oxidation states assigned on both sides of the equation for atoms of similar chemical identity, the actual magnitude of change in oxidation state for the atom of each of the reacting species on the left-hand side of the equation. Again, the magnitude of change refers to that of the single atom.

3. Determine and record the change in oxidation number for the affected atoms on the left. This involves merely multiplication of the oxidation state of the particular atom, as previously determined, by the numbers of such atom present in the entire chemical formula unit.

4. Equalize the loss and gain of electrons denoted by the foregoing multiplication by *whole numbers*, to provide the required ratio of reacting formula units. This ensures that the net total decrease in oxidation numbers of all formula units of the oxidant equals the net total increase in oxidation numbers of all formula units of the reductant. These whole numbers now represent the coefficients of the redox reactants on the left. These coefficients are conventionally reduced to the simplest empirical ratio when they happen to have a common divisor — although nonetheless still correct even as a multiple, inasmuch as the coefficients signify merely the relative numbers of reacting species and not their absolute quantities.

5. Equalize the number of atoms of the redoxed species on the right of the equation with their counterparts now on the left, by supplying the correct coefficients to the right. In the frame now established by the redox coefficients on both sides, balance by inspection all remaining substances present — both left and right, in conformity with the chemically-identical equal numbers of atoms demanded by the law of convervation of mass.

These coefficients, as derived and as assigned both on the left and on the right, must not be changed except when part of a species has been used, not in redox but in metathetical (nonredox) association. The arithmetical additions of these metathetical formula units to those of the established redox formula units, on the relatively few occasions when they are encountered, are always deferred until after the appropriate redox coefficients have been assigned on both sides of the equation. These additions offer no difficulties; they are readily apparent by inspection once redox requirements are known.

BALANCING OF OXIDATION NUMBERS (ALL PRODUCTS GIVEN)

The illustrations that follow, for the sake of clarity in developing the ultimate stoichiometrically correct and balanced equation from the given skeletal (unbalanced) equation, portray the step-by-step progress toward the finished equation by means of appropriate notational references. The student should carefully follow these steps in their proper sequence, as delineated by the lettering (a), (b), (c), (d), and (e), respectively, in order to ensure a complete grasp of the balancing mechanism. The charges assigned appear appropriately both directly above and below the affected atoms.

A. Molecular Stoichiometry; Nondisproportionation

■ EXAMPLE 1:   Potassium dichromate + ferrous sulfate in sulfuric acid medium.

(a) Oxidation state (skeletal equation):

$$\overset{+6}{K_2Cr_2O_7} + \overset{+2}{FeSO_4} + H_2SO_4 \rightarrow K_2SO_4 + \overset{+3}{Cr_2(SO_4)_3} + \overset{+3}{Fe_2(SO_4)_3} + H_2O$$

(b) Change in oxidation state:

Cr 3 units $(+6 \rightarrow +3)$ | Fe 1 unit $(+2 \rightarrow +3)$

(c) Change in oxidation number:

Cr 6 units (2 atoms $\times$ 3/atom) | Fe 1 unit (1 atom $\times$ 1/atom)

(d) Multiples for formulas: 1 $K_2Cr_2O_7$ | 6 $FeSO_4$

(e) Balanced equation:

$$K_2Cr_2O_7 + 6FeSO_4 + 7H_2SO_4 \rightarrow K_2SO_4 + Cr_2(SO_4)_3$$
$$+ 3Fe_2(SO_4)_3 + 7H_2O.$$

Had the presence of excess H_2SO_4 favored the formation of the potassium acid sulfate, $KHSO_4$, the balanced equation for such reaction would read:

$$K_2Cr_2O_7 + 6FeSO_4 + 8H_2SO_4 \rightarrow 2KHSO_4$$
$$+ Cr_2(SO_4)_3 + 3Fe_2(SO_4)_3 + 7H_2O.$$

The molecular equation may be transcribed to ionic stoichiometry simply by rewriting all wholly or largely electrovalent compounds in the forms of their independently existent and mobile (in aqueous medium) constituent ions. Retained in the molecular notation are all species that are predominantly covalent and weakly ionized; and likewise retained in molecular notation are precipitated or insoluble electrolytes, as a practical expression of their lack of ionic mobility in the solid state. Care must be exercised during this conversion to maintain accurately the equivalences of atomic and electronic charges, as already established, particularly in the process of excluding from both sides of the ionic equation all species that are not chemically altered.

In this connection we must bear in mind that the covalent bonding of two or more originally independent ions (or their formation by ionization from a parent covalent species) constitutes an alteration in their chemical behavior; hence, they must be written to conform to the consistencies inherent in a written equation for an ionic reaction. The ionic equation for the given reaction that has just been molecularly developed thus becomes:

$$Cr_2O_7^{2-} + 6Fe^{2+} + 14H^+ \rightarrow 2Cr^{3+} + 6Fe^{3+} + 7H_2O.$$

A prudent safeguard in this conversion is always to check upon the equality of net ionic charge of all species on each side of the ionic equation; proved here by a net of 24+ for reactants and 24+ for products.

■ EXAMPLE 2:   Potassium permanganate + phosphine in sulfuric acid medium.

(a) Oxidation state (skeletal equation):

$$\overset{+7}{K}MnO_4 + \overset{-3}{P}H_3 + H_2SO_4 \rightarrow K_2SO_4 + \overset{+2}{Mn}SO_4 + \overset{+5}{H_3P}O_4 + H_2O$$

(b) Change in oxidation state:

Mn 5 units $(+7 \rightarrow +2)$ | P 8 units $(-3 \rightarrow +5)$

(c) Change in oxidation number:

Mn 5 units $(1 \text{ atm} \times 5/\text{atm})$ | P 8 units $(1 \text{ atm} \times 8/\text{atm})$

(d) Multiples for formulas: 8 $KMnO_4$ | 5 PH_3

(e) Balanced equation:

$$8KMnO_4 + 5PH_3 + 12H_2SO_4 \rightarrow 4K_2SO_4 + 8MnSO_4$$
$$+ 5H_3PO_4 + 12H_2O$$

and converted, if required, to ionic stoichiometry:

$$8MnO_4^- + 5PH_3 + 24H^+ \rightarrow 8Mn^{2+} + 5H_3PO_4 + 12H_2O.$$

This equation balances with a net ionic charge of $16+$ on each side.

B. Ionic Stoichiometry; Nondisproportionation

■ EXAMPLE 1: Metallic zinc + dilute nitric acid.

(a) Oxidation state (skeletal equation):

$$\overset{0}{Zn}{}^0 + \overset{+5}{N}O_3^- + H^+ \rightarrow \overset{+2}{Zn}{}^{2+} + \overset{-3}{N}H_4^+ + H_2O$$

(b) Change in oxidation state:

Zn 2 units $(0 \rightarrow +2)$ | N 8 units $(+5 \rightarrow -3)$

(c) Change in oxidation number:

Zn 2 units $(1 \text{ atm} \times 2/\text{atm})$ | N 8 units $(1 \text{ atm} \times 8/\text{atm})$

(d) Multiples for formulas: 4 Zn^0 | 1 NO_3^-

(e) Balanced equation:

$$4Zn^0 + NO_3^- + 10H^+ \rightarrow 4Zn^{2+} + NH_4^+ + 3H_2O.$$

This ionic equation tells the entire story of the chemical change. It balances with equal net ionic charges of $9+$ on each side. Transcribing to molecular stoichiometry, when desired, requires, however, a knowledge of the sources of supply of both the NO_3^- and H^+ ions and brings into important consideration the additional positive and negative ions needed to associate metathetically with the ions on both sides of the equation. Remembering that all molecules are electrically neutral (as a net), eight additional NO_3^- ions are required to associate (to crystallize $Zn(NO_3)_2$ from solution) with the four already stoichiometrically established as present. Further, one more NO_3^- ion is required to crystallize the one NH_4^+ ion likewise established in the redox balancing. Altogether this makes nine extra

NO_3^- ions that must be added to the one NO_3^- ion that has already been calculated as undergoing reduction, or a total of ten. As the source of supply of both the NO_3^- and H^+ ions is the HNO_3, the molecular equation then becomes

$$4Zn^0 + 10HNO_3 \rightarrow 4Zn(NO_3)_2 + NH_4NO_3 + 3H_2O.$$

For the $10H^+$ in this equation there must also be precisely $10NO_3^-$, because both ions are produced from the HNO_3 in equal numbers. The apparent discrepancy of the missing $9NO_3^-$ in the balanced ionic equation is now reconciled through awareness that the ionic equation depicts the identities and stoichiometric magnitudes solely of those species that actually undergo chemical change. In this particular reaction, only one out of every ten NO_3^- ions is chemically used up by reduction to NH_4^+ ion. The other nine still remain unaltered but they are present to maintain total over-all electrical neutrality of the dissolved substances.

In further illustration of the metathetical involvements in the conversion of an ionic redox equation to molecular form, let us consider the reaction between metallic copper and dilute nitric acid, for which the balanced ionic redox equation is:

$$3Cu^0 + 2NO_3^- + 8H^+ \rightarrow 3Cu^{2+} + 2NO^0 + 4H_2O.$$

Its molecular transcription into

$$3Cu^0 + 8HNO_3 \rightarrow 3Cu(NO_3)_2 + 2NO^0 + 4H_2O$$

confirms that only two of the eight moles of HNO_3 are used in the reaction to form NO^0. The other six must supply the necessary balance of over-all electrical neutrality of the reaction system as a whole, in terms of metathetical ionic associations.

■ EXAMPLE 2:   Potassium permanganate + hydrogen peroxide in sulfuric acid medium.

(a) Oxidation state (skeletal equation):

$$\overset{+7}{MnO_4^-} + \overset{-1}{H_2O_2} + H^+ \rightarrow \overset{+2}{Mn^{2+}} + H_2O + \overset{0}{O_2}$$

(b) Change in oxidation state:
    Mn 5 units $(+7 \rightarrow +2)$ | O 1 unit $(-1 \rightarrow 0)$

(c) Change in oxidation number:
    Mn 5 units (1 atm × 5/atm) | O 2 units (2 atms × 1/atm)

(d) Multiples for formulas: 2 MnO_4^- | 5 H_2O_2

(e) Balanced equation:

$$2MnO_4^- + 5H_2O_2 + 6H^+ \rightarrow 2Mn^{2+} + 8H_2O + 5O_2.$$

Converting to molecular stoichiometry, if desired, yields

$$2KMnO_4 + 5H_2O_2 + 3H_2SO_4 \rightarrow K_2SO_4 + 2MnSO_4 + 8H_2O + 5O_2$$

or, if excess of H_2SO_4 should produce the acid salt of potassium,

$$2KMnO_4 + 5H_2O_2 + 4H_2SO_4 \rightarrow 2KHSO_4 + 2MnSO_4 + 8H_2O + 5O_2.$$

In the ionic equation, note that the particular oxygen species on the right, which represents the conversion product of the peroxide, must be chosen carefully. The peroxy-oxygen atom is quite versatile for use in experiments. Under variable and suitably favorable conditions it may function both as an oxidant, changing in oxidation state from -1 to -2, and as a reductant, changing from -1 to 0. Note, too, that the arithmetical (not experimental) opportunities with respect to such duality of role present themselves in the "bookkeeping" of the products. As the MnO_4^- ion is being reduced, it is inevitable that the H_2O_2 must be oxidized; hence, the product of its change is not the H_2O but, again inevitably, the O_2.

Upon this critically important appreciation of the concept that there can be no reduction without accompanying oxidation depends the correct assignments of the coefficients for the products of reaction and the validity of the written equation. Here, therefore, the oxygen atoms from the oxidation of the peroxide must be balanced in the O_2 and not in the H_2O. Any other course results in completely incorrect stoichiometry, because it is easily possible to assign incorrect coefficients to the products and still obtain equal numbers of atoms on both sides. The rank inaccuracy in such an error is readily manifest in the failure of the net ionic charge to balance on both sides of the equation. A check upon the balance of net ionic charges is always quick, convenient, and sure when ionic equations are evaluated, but, regrettably, it does not offer itself with molecular equations. In such molecular stoichiometry, it is always possible to total algebraically the oxidation numbers of all the species on the left and check for equality with those on the right. This, however, is a quite laborious chore when reactions are complex.

In the foregoing reaction it cannot be the oxygen atoms of the MnO_4^- ion that are being oxidized to O_2 gas. The solubilization of $KMnO_4$ alone in a dilute water solution of sulfuric acid produces no chemical effects whatever of this nature; nor are there any other significant alterations in the chemical identity of either solute or solvent in such a solution.

C. Simple and Multiple Disproportionation

In disproportionation ("self-," "internal-," "autonomous redox") a single chemical species undergoes *both* increase and decrease in oxidation state. That is, it loses and gains electrons simultaneously, certain of the formula units of the disproportionating species yielding the electrons for seizure by the remaining units. When only one reaction product represents oxidation and only one, likewise, represents reduction, the disproportionation is called "simple." When two or more reaction products are encountered separately either in oxidation or in reduction, the disproportionation is called "multiple."

The equation-balancing procedure here is precisely the same as that employed when oxidant and reductant are chemically different. It is most convenient to regard the oxidant and reductant units of the disproportionating species as being mutually separated and, then establish in conventional fashion their coefficients of interaction, merely adding the separate but chemically identical formula units. The following examples will serve in illustration:

■ EXAMPLE 1:   The simple disproportionation of manganate ion in acidic medium.

(a) Oxidation state (skeletal equation):

$$\overset{+6}{MnO_4^{2-}} + H^+ \rightarrow \overset{+4}{MnO_2} + \overset{+7}{MnO_4^-} + H_2O$$

(b) Change in oxidation state:
    Mn 2 units $(+6 \rightarrow +4)$| Mn 1 unit $(+6 \rightarrow +7)$
(c) Change in oxidation number:
    Mn 2 units $(1 \text{ atm} \times 2/\text{atm})$| Mn 1 unit $(1 \text{ atm} \times 1/\text{atm})$
(d) Multiples for formulas: 1 MnO_4^-| 2 MnO_4^-
(e) Balanced equation:

$$3MnO_4^{2-} + 4H^+ \rightarrow MnO_2 + 2MnO_4^- + 2H_2O.$$

■ EXAMPLE 2:   The multiple disproportionation of sodium monohydrogen phosphite ignited in dry state:

(a) Oxidation state (skeletal equation):

$$\overset{+3}{Na_2HPO_3} \rightarrow \overset{+5}{Na_3PO_4} + \overset{+5}{Na_4P_2O_7} + \overset{-3}{PH_3} + H_2O.$$

It is observed that although there is only one product of reduction — phosphine PH_3 — there are two products of oxidation, namely, trisodium phosphate, Na_3PO_4, and sodium pyrophosphate, $Na_4P_2O_7$. The balancing procedure here, quite logically, is to break down the disproportionation conveniently into two separate but simultaneous oxidation-reduction processes, and then, as they are chemically identical, merely to total the chemical units of the Na_2HPO_3 required for each. Hence, continuing in this theme:

Redox A

$$\overset{+3}{\underset{|}{Na_2HPO_3}} \xrightarrow{\text{redox}} \overset{+5}{\underset{|}{Na_4P_2O_7}} + \overset{-3}{\underset{|}{PH_3}}$$

(b) Change in
oxidation state: P 2 units $(+3 \rightarrow +5)$ | P 6 units $(+3 \rightarrow -3)$

(c) Change in
oxidation number: P 2 units $\left(1 \text{ atm} \times \dfrac{2}{\text{atm}}\right)$ | 6 units $\left(1 \text{ atm} \times \dfrac{6}{\text{atm}}\right)$

(d) Multiples for formulas: 3 Na_2HPO_3 | 1 Na_2HPO_3

Redox B

$$\overset{+3}{\underset{|}{Na_2HPO_3}} \xrightarrow{\text{redox}} \overset{+5}{\underset{|}{Na_3PO_4}} + \overset{-3}{\underset{|}{PH_3}}$$

(b) Change in
oxidation state: P 2 units $(+3 \rightarrow +5)$ | P 6 units $(+3 \rightarrow -3)$

(c) Change in
oxidation number: P 2 units $\left(1 \text{ atm} \times \dfrac{2}{\text{atm}}\right)$ | 6 units $\left(1 \text{ atm} \times \dfrac{6}{\text{atm}}\right)$

(d) Multiples for formulas: 3 Na_2HPO_3 | 1 Na_2HPO_3.

There are then, *altogether*, eight formula units of the Na_2HPO_3 required; a total of four each for redox A and for redox B. Of this total of eight, two must be reduced to PH_3 and six must be oxidized to proper conversions of phosphorus atoms within Na_3PO_4 and $Na_4P_2O_7$. A judicious inspection of the skeletal equation in the light of the redox coefficients that have been developed lead to the assigning of correct coefficients to these two different species of oxidation products. As one approach in appraisal, it is seen that the coefficient for the $Na_4P_2O_7$ must be less than 3 if the phosphorus atoms required for the product, Na_3PO_4, are to be available. *If* the coefficient to be assigned to the $Na_4P_2O_7$ were 2, there would have to be two formula units of Na_3PO_4 for balance of the phosphorus atoms. As this would leave the sodium atoms unbalanced, such an assignment is unacceptable. If, on the other hand, the coefficient for the $Na_4P_2O_7$ were 1, that of the Na_3PO_4 would obviously be 4. That this is now correct is proved by the accurate balancing of the sodium atoms — 16 on each side. Therefore, the valid result conforms to

(e) Balanced equation:
$8Na_2HPO_3 \rightarrow 4Na_3PO_4 + Na_4P_2O_7 + 2PH_3 + H_2O.$

Multiple disproportionations of even broader magnitudes may be handled in similar fashion by the method of balancing oxidation numbers.

DERIVING FULL REDOX EQUATION FROM PARTIAL PRODUCTS

As by far the greatest number of reactions with electrolytes occur in aqueous solution, a complete assessment of the method of balancing by oxidation numbers should include consideration of whether, as with the ion-electron method, shortly to be developed, it is possible to work out a completely balanced equation without being given *all* of the reaction products at the start. In either method the equation-balancer must know the formulas of *all* reactants and reaction products in which changes in oxidation state occur if the result obtained is to conform correctly to the stoichiometry of actual experiments.

It is possible, however, to apply with the method of oxidation numbers the comparable device of the ion–electron method, which permits non-redoxed species emanating directly from the solvent to be worked into the equation in systematic and progressive fashion, as required during the balancing process, and as compatible with their stabilities in the chemical environment of the given reaction.

Thus, in an ionic reaction in aqueous acidic medium, H^+ ion may be inserted directly into the equation to supply not only hydrogen atoms where and when needed, but also positive charges to the side deficient therein, or to the side requiring diminution of its excess net ionic negative charge for equalization. Where oxygen atoms are required for balancing purposes in an acidic medium of an aqueous solution, H_2O molecules are abundantly available for such needs. In an alkali-basic medium, such as $NaOH$ or KOH, abundant OH^- ion is available to supply not only oxygen and hydrogen atoms but also negative charges to the side deficient therein or to the side that needs diminution of excess ionic positive charge. H^+ may not be used here to supply hydrogen atoms because, as a result of the neutralization of H^+ ion, no significant quantities exist.

Neutral H_2O molecules are always abundantly available to supply both hydrogen and oxygen atoms when these must be added without disrupting net ionic charge. In ammonia-basic (ammoniacal) solution, where OH^- ion is present in only limited amounts, NH_4^+ ion is available to supply requisite ionic positive charge to the side of the reaction that is deficient therein; or, as needed, to the side that requires the diminution of net negative charge. In an ammoniacal medium, NH_3 molecules are available for the balancing of nitrogen atoms; and, as in all aqueous media, regardless of the nature or extent of the acid or base present, H_2O molecules for the balancing of any oxygen atoms.

COMPLETING PARTIAL SKELETAL EQUATION

The following simple rules, to be followed in the exact sequence given, are of advantage in conveniently completing partial skeletal equations of ionic redox reactions that are to be balanced by oxidation numbers, through a systematic plan of introducing the necessary species of solvent.

1. Determine and assign the correct redox coefficients to both sides of the equation in the accustomed manner.

2. Check the net ionic charges on both sides. To balance the net ionic charge (electrical): If the solution is *acidic*, supply H^+; if *strongly alkali-basic*, with NaOH or KOH, use OH^-; if *ammoniacal*, use NH_4^+.

3. After net ionic charge has been equalized on both sides of the equation by inserting the aforementioned pertinent ions on the appropriate side, check the reactions of *acidic* and *alkali-basic* media with respect to oxygen atoms and, wherever they are deficient, introduce H_2O.

This step provides simultaneously not only the oxygen atoms, but also the requisite balancing hydrogen atoms that were introduced by H^+, in acidic medium and OH^- in basic medium in the preceding step for the purpose of equalizing ionic charges for which compensation must be made. These ions should, be counted as a final check on the correctness of the equation. In ammoniacal medium, balance the nitrogen atoms introduced by NH_4^+ during the equalizing of ionic charges by adding the requisite number of NH_3 molecules to the appropriate side. Then balance any oxygen atoms by introducing H_2O molecules, where needed. The hydrogen atoms likewise balance upon completion of these processes. The accuracy of the final stoichiometry should be given a final check by ascertaining that their numbers are equal on both sides of the equation.

A few words are due with respect to the utilization of NH_4^+ to equalize net ionic charge in ammoniacal medium, as it would seem just as valid, arithmetically, to insert OH^- on the side with excess of positive charge for the same purpose. Certainly, the NH_4^+ would have no numerical advantage over the OH^- because, in the absence of added common ion, or of chemical destruction or removal of either ion, the conditions for the equilibrium

$$(NH_3 + H_2O \rightleftarrows NH_4^+ + OH^-, \text{ or } NH_4OH \rightleftarrows NH_4^+ + OH^-)$$

must, perforce, produce and maintain both of these ions in solution in equal numbers.

There is no refutation here of the experimental observation that, in equivalent quantities, OH^- ion functions as a stronger base than NH_3 (NH_4OH). In fact, OH^- ion is the strongest base that can exist in aqueous medium, because a base stronger than OH^- ion, when introduced into water, must necessarily dissipate itself in competitive reaction for protons (available from H_2O) with consequent formation of OH^- ion. Thus, using the stronger base, NH_2^- (the amide ion), as an example, the reaction is

$$NH_2^- + H_2O \rightleftarrows NH_3 + OH^-.$$

The introduction of NH_4^+ and NH_3 for balancing the equations of reactions in ammoniacal medium merely emphasizes the importance of concentration in the inherent capacity of a substance to function in homogeneous reaction.

In this type of environment, wherein the concentration of molecular NH_3 is very high, and the concentration of the OH^- ion is extremely low, the NH_3 may, for practical purposes, be regarded as the prime basic species initially present. Consequently, it will prove the more effective neutralizing agent to H^+ ion under the given conditions.

BALANCING OF OXIDATION NUMBERS (PRODUCTS PARTLY GIVEN)

The foregoing considerations, as applied to the three common reaction media, are now illustrated by the following specific examples. Their stepwise sequential developments should be very carefully pursued.

■ EXAMPLE 1. *In acidic medium:*   Dichromate ion + oxalic acid + dilute sulfuric acid.

(a) Oxidation state (partial skeletal equation):

$$\overset{+6}{Cr_2O_7^{2-}} + \overset{+3}{H_2C_2O_4} \rightarrow \overset{+3}{Cr^{3+}} + \overset{+4}{CO_2}$$

(b) Change in oxidation state:

Cr 3 units $(+6 \rightarrow +3)$ | C 1 unit $(+3 \rightarrow +4)$

(c) Change in oxidation number:

Cr 6 units (2 atm $\times$ 3/atm) | C 2 units (2 atms $\times$ 1/atm)

(d) Multiples for formulas: 1 $Cr_2O_7^{2-}$ | 3 $H_2C_2O_4$

(e) Balanced equation (progressive):

(1) Assignment of redox coefficients:

$$Cr_2O_7^{2-} + 3H_2C_2O_4 \rightarrow 2Cr^{3+} + 6CO_2.$$

(2) Assignment of equalizing ionic charge, by H^+:

$$Cr_2O_7^{2-} + 3H_2C_2O_4 + 8H^+ \rightarrow 2Cr^{3+} + 6CO_2.$$

(3) Assignment of H_2O to balance oxygen atoms:

$$Cr_2O_7^{2-} + 3H_2C_2O_4 + 8H^+ \rightarrow 2Cr^{3+} + 6CO_2 + 7H_2O.$$
(the finished equation)

■ EXAMPLE 2. *In alkali-basic medium* (NaOH, KOH):   metallic zinc + nitrite ion + sodium hydroxide.

(a) Oxidation state (partial skeletal equation):

$$\overset{0}{Zn^0} + \overset{+3}{NO_2^-} \rightarrow \overset{+2}{ZnO_2^{2-}} + \overset{-3}{NH_3}$$

(b) Change in oxidation state:

$$\text{Zn} \quad 2 \text{ units } (0 \to +2) \mid \text{N} \quad 6 \text{ units } (+3 \to -3)$$

(c) Change in oxidation number:

$$\text{Zn} \quad 2 \text{ units } (1 \text{ atm} \times 2/\text{atm}) \mid \text{N} \quad 6 \text{ units } (1 \text{ atm} \times 6/\text{atm})$$

(d) Multiples for formulas: $3 \text{ Zn}^0 \mid 1 \text{ NO}_2^-$

(e) Balanced equation (progressive):

(1) Assignment of redox coefficients:

$$3\text{Zn}^0 + \text{NO}_2^- + 5\text{OH}^- \to 3\text{ZnO}_2^{2-} + \text{NH}_3 + \text{H}_2\text{O}$$

(2) Assignment of equalizing ionic charge by OH^-:

$$3\text{Zn}^0 + \text{NO}_2^- + 5\text{OH}^- \to 3\text{ZnO}_2^{2-} + \text{NH}_3$$

(3) Assignment of H_2O to balance oxygen atoms:

$$3\text{Zn}^0 + \text{NO}_2^- + 5\text{OH}^- \to 3\text{ZnO}_2^{2-} + \text{NH}_3 + \text{H}_2\text{O}.$$
(the finished equation)

■ EXAMPLE 3. In *ammonia-basic (ammoniacal) medium.*　　Bromine + aqueous ammonia.

(a) Oxidation state (partial skeletal equation):

$$\overset{0}{\text{Br}_2} + \overset{-3}{\text{NH}_3} \to \overset{-1}{\text{Br}^-} + \overset{0}{\text{N}_2}$$

(b) Change in oxidation state:

$$\text{Br} \quad 1 \text{ unit } (0 \to -1) \mid \text{N} \quad 3 \text{ units } (-3 \to 0)$$

(c) Change in oxidation number:

$$\text{Br} \quad 2 \text{ units } (2 \text{ atms} \times 1/\text{atm}) \mid \text{N} \quad 3 \text{ units } (1 \text{ atm} \times 3/\text{atm})$$

(d) Multiples for formulas: $3 \text{ Br}_2 \mid 2 \text{ NH}_3$

(e) Balanced equation (progressive):

(1) Assignment of redox coefficients:

$$3\text{Br}_2 + 2\text{NH}_3 \to 6\text{Br}^- + \text{N}_2.$$

(2) Assignment of equalizing ionic charge, by NH_4^+:

$$3\text{Br}_2 + 2\text{NH}_3 \to 6\text{Br}^- + \text{N}_2 + 6\text{NH}_4^+.$$

(3) Assignment of NH_3 to balance nitrogen atoms:

$$3\text{Br}_2 + 8\text{NH}_3 \to 6\text{Br}^- + \text{N}_2 + 6\text{NH}_4^+$$
i.e., $2\text{NH}_3 + 6\text{NH}_3$ additional
(the finished equation)

In conversion to *molecular* stoichiometry, if sought, and as already amply described, we obtain:

$$3\text{Br}_2 + 8\text{NH}_3 \to 6\text{NH}_4\text{Br} + \text{N}_2.$$

Were it preferred to represent aqueous NH_3 as NH_4OH, the ionic equation for the previous reaction would have been developed as follows:

(1) Assignment of redox coefficients:

$$3Br_2 + 2NH_4OH \rightarrow 6Br^- + N_2.$$

(2) Assignment of equalizing ionic charge, by NH_4^+:

$$3Br_2 + 2NH_4OH \rightarrow 6Br^- + N_2 + 6NH_4^+.$$

(3) Assignment of NH_4OH to balance nitrogen atoms:

$$3Br_2 + 8NH_4OH \rightarrow 6Br^- + N_2 + 6NH_4^+.$$
i.e., $2NH_4OH + 6NH_4OH$ additional

(4) Assignment of H_2O to balance oxygen atoms:

$$3Br_2 + 8NH_4OH \rightarrow 6Br^- + N_2 + 6NH_4^+ + 8H_2O.$$
(*the finished equation*)

■ EXAMPLE 4. *In ammonia-basic medium:*   Arsenic sulfide + hydrogen peroxide.

(a) Oxidation state (partial skeletal equation):

$$\overset{-2}{\underset{}{As_2}}\overset{-1}{S_5} + \overset{}{H_2}\overset{-1}{O_2} \rightarrow \overset{+6}{As}O_4^{3-} + \overset{-2}{S}O_4^{2-} + H_2O$$

(b) Change in oxidation state:
S 8 units $(-2 \rightarrow +6)$ | O 1 unit $(-1 \rightarrow -2)$

(c) Change in oxidation number:
S 40 units (5 atm × 8/atm) | O 2 units (2 atms × 1/atm)

(d) Multiples for formulas: 1 As_2S_5 | 20 H_2O_2

(e) Balanced equation (progressive):

(1) Assignment of redox coefficients:

$$As_2S_5 + 20H_2O_2 \rightarrow 2AsO_4^{3-} + 5SO_4^{2-} + 40H_2O.$$

(2) Assignment of equalizing ionic charge by NH_4^+:

$$As_2S_5 + 20H_2O_2 \rightarrow 2AsO_4^{3-} + 5SO_4^{2-} + 40H_2O + 16NH_4^+.$$

(3) Assignment of NH_3 to balance nitrogen atoms:

$$As_2S_5 + 20H_2O_2 + 16NH_3 \rightarrow 2AsO_4^{3-} + 5SO_4^{2-} + 40H_2O + 16NH_4^+.$$

(4) Assignment of H_2O to balance oxygen atoms:

$$As_2S_5 + 20H_2O_2 + 16NH_3 + 28H_2O \rightarrow 2AsO_4^{3-} + 5SO_4^{2-} + 40H_2O + 16NH_4^+.$$

As water molecules on both sides may be algebraically combined, the ultimate refinement then reads:

$$As_2S_5 + 20H_2O_2 + 16NH_3 \rightarrow 2AsO_4^{3-} + 5SO_4^{2-} + 12H_2O + 16NH_4^+.$$
(the finished equation)

A final check reveals 88 hydrogen atoms on each side of the equation.

In the event that formulation of aqueous ammonia as NH_4OH be preferred, the development of the final balanced equation would have been progressively as follows:

(1) Assignments of redox coefficients and of NH_4^+ to equalize ionic charge:

$$As_2S_5 + 20H_2O_2 \rightarrow 2AsO_4^{3-} + 5SO_4^{2-} + 40H_2O + 16NH_4^+.$$

(2) Assignment of NH_4OH to balance nitrogen atoms:

$$As_2S_5 + 20H_2O_2 + 16NH_4OH \rightarrow 2AsO_4^{3-} + 5SO_4^{2-} + 40H_2O + 16NH_4^+.$$

(3) Assignment of H_2O to balance oxygen atoms:

$$As_2S_5 + 20H_2O_2 + 16NH_4OH + 12H_2O \rightarrow 2AsO_4^{3-}$$
$$+ 5SO_4^{2-} + 40H_2O + 16NH_4^+.$$

Upon algebraic combination of water molecules appearing on both sides, this reduces to:

$$As_2S_5 + 20H_2O_2 + 16NH_4OH \rightarrow 2AsO_4^{3-} + 5SO_4^{2-} + 28H_2O + 16NH_4^+.$$
(the finished equation)

Again, the numbers of hydrogen atoms on both sides of the equation are 120 on left, 120 on right.

BALANCING BY ION–ELECTRON METHOD: HALF-REACTIONS

The judicious intent in the rather intimate attention that has been given to the method of balancing equations by oxidation numbers has been to stress the broad applicability of this method to all types of reactions and elucidated mechanisms, regardless of the homogeneity or heterogeneity of the system's phases. Indeed, the objective has been to demonstrate that the method is completely indispensable to the balancing of equations for reactions occurring in nonaqueous media, whether gaseous, liquid, or fused solid. We have also emphasized the validity of the opportunities afforded in aqueous media to elucidate reaction products emanating from the solvent itself, when such products were not initially given as part of the skeletal equation.

The considerations that have been developed must not be allowed to distract, let alone detract, from the essential merits of the *ion–electron method* (or *method of half-reactions*) of balancing equations of reactions occurring in aqueous media. This particular method is rather exclusively restricted to such aqueous environments as a specialized approach to recognizing and interpreting the chemical behavior of ionic species in a common, conventional solvent conducive to their independent existence in mutual equilibria with one another.

Not that the ion–electron method is necessarily surer or even faster than the method of oxidation numbers when applied to aqueous systems. In fact, comparisons often favor the latter in point of view of time, when reaction products are rather complex. In multiple disproportionations, however, involving the production of several oxidation or reduction products, the advantage in facility is definitely with the ion–electron method. In such instances the oxidation number method normally requires extended development in several steps, as was shown in the illustrations of this particular process given in the preceding section.

The special merits of the ion–electron method of balancing are revealed, rather, in the concomitant opportunities afforded in balancing an equation to interpret the most logical mechanisms of reaction. In many instances, to be sure, these appraisals are hardly more than plausible elucidations, not readily susceptible (and frequently not even possible) of experimental proof. The ion–electron method provides this more logical interpretative frame by completely eliminating from needed consideration the highly arbitrary, and frequently quite unreal, assignments of oxidation state in the oxidation number method. The ion–electron method contends not with apparent charges on species but with the actual charges of species that exist independently in the ionizing atmosphere of the solvent.

Thus, the polyatomic permanganate ion, MnO_4^-, represents a monovalent charged collective unit, not only for purposes of balancing equations, but also for appreciation of its properties as an independently existing and indivisible chemical species when so introduced into water solution. When the MnO_4^- ion is deemed to function as an oxidant no suggestion is made that, in reduction to Mn^{2+} or MnO_2 (among others), it is the manganese atom alone in the entire polyatomic species that does the oxidizing, and that, consequently, it alone acquires the electrons rather than the polyatomic ion as a whole.

As previously noted, this experimentally untenable suggestion that the manganese atom alone may be exclusively singled out for attention represents the concept of the over-simplified atomic orbital theory that cannot be proved, but which represents the arithmetically secure basis for assigning of oxidation states. Were it not for the present desirability of maintaining elementary pedagogical approaches on a level consistent with practical expediency in a valid mathematical frame, certainly the substitution of atomic orbital concepts for the more realistic (albeit, far more complicated and difficult) portrayals of the molecular orbital concepts could hardly be invited, or even reconciled.

The ion–electron half-reactions (or "partials" as they are frequently termed), which represent the separate oxidation and reduction steps leading to the over-all net equation of redox, provide rewarding opportunities to evaluate, not only the probability that a reaction will occur spontaneously under normal conditions of temperature, concentrations of reactant, and relative pH of the solution, but also, when other information is supplied, even the magnitude or extent of reaction in the direction depicted by the net equation. These appraisals are forthcoming either from directly-available oxidation potentials or oxidation constants for the half-reactions, as written, or by their calculations from data provided. What must be appreciated here, as an indispensable adjunct of the correctly balanced equation, *per se*, is the applicability of the equation thereafter to the calculations in which it will be employed. It is in this latter area of interpretation, wherein the chemical identity and behavior of species in ionic aqueous environment becomes important, that the ion-electron method shows its *superior* advantages.

RULES FOR SYSTEMATIC BALANCING

We now proceed to the actual stepwise mechanism by which redox reactions in water solution may be given net over-all portrayal in correct ionic and molecular stoichiometry. This sequence of systematic development should be carefully followed:

1. *Set up skeletal half-reactions of oxidation and reduction.*

These ion–electron "partials" are individually separate equations depicting the separate behavior of oxidant and reductant. Select either one for the first half-reaction, balance it completely in careful conformity with stated procedure, and then do the same with the second "partial."

(a) Start half-reaction by writing selected species in precisely the form that represents the predominant character of its initial existence and availability in the reaction medium.

This is as an ion (simple or polyatomic) if its association with other species is predominantly electrovalent (characteristic of true salts, either solid or dissolved in aqueous solution: as a molecule, if covalent (normal or coordinate) bonding predominates to yield a net uncharged species. A weakly ionized species, therefore, is to be written molecularly, unless it is a weakly dissociated complex ion which, as a collective unit, is electrovalently associated with other ionic species (as in complex salts), and which, as such, is to be represented ionically as a charged polyatomic complex.

In the molecular category are placed weak acids and weak bases (including H_2O itself), all gaseous species (dissolved or evolved), and also solids as a concession to the lack of mobility of ions associated in a precipitated or difficultly soluble substance. The completion of the correct skeletal equation is a happy consequence of both mental restraint and understanding of the chemistry involved. The restraint is in avoiding the temptation to introduce any other reactants, no matter how obvious, before the next step.)

(b) Draw the arrow, allowing adequate space for the subsequent introduction of additional reactants (as is progressively delineated), and then add the specific analogue(s) of redox.

These conform to the starting species, reduction product of the oxidant, or oxidation product of the reductant in the reaction. Again, as already stressed, the species so introduced at the right of the arrow (the products) must conform to the ionic or molecular delineations representative of their predominant character in the reaction medium.

2. *Balance equation atomically.*

(a) *Atoms other than* O *and* H: With the oxidized and reduced analogues of the selected species now stated on both sides of the arrow, insert appropriate coefficients to balance the numbers of all atoms other than those of oxygen and hydrogen. These are supplied shortly, depending on the nature of the aqueous medium; that is, whether it is acidic, alkali-basic, or ammonia-basic.

Before balancing O and H, draw upon any species available in the system — in conformity with their precise identities (actually given or learned by experiment), as may be required for balancing other atoms.

(b) *Oxygen atoms:* Check the oxygen atoms on both sides of the equation and insert the required number on the side deficient therein as follows:

(1) For *acidic* or *ammoniacal* reaction media: Use H_2O molecules in requisite numbers.

(2) For *alkali-basic* media: As the concentration of OH^- ion is relatively very high (from NaOH, KOH), insert OH^-, itself, in requisite numbers on the side deficient in oxygen atoms. This is the most direct means for balancing oxygen atoms.

It is equally correct here to supply oxygen atoms by introducing molecules of H_2O because both OH^- and H_2O represent abundant and completely realistic sources of supply of oxygen atoms in an alkali-basic medium. The preference for OH^- as the representative source of supply, however, is dictated by the anticipation that any excess of H^+ ion to be met shortly must inevitably call for the introduction of neutralizing OH^- in the alkali-basic medium which results in the formation of H_2O. This is because any maintenance of H^+ ion in a chemical equation for a reaction in this medium would unmistakably denote significant amounts. Hence it is prohibited inasmuch as significant quantities of H^+ ion in the solution would be completely incompatible with the alkaline environment of the reaction. Moreover, as the hydrogen atoms in alkaline medium are to be balanced by addition of H_2O (that is, H—OH) to the side deficient in hydrogen atoms, with the acompanying counterbalancing of permissible OH^- on the opposite side, the eventual collation of water molecules appearing on both sides of the equation involve an additional step in the conventional simplification of the equation.

(c) *Hydrogen atoms:* Check the hydrogen atoms on both sides of the equation, and balance them as follows in accordance with the nature of the particular chemical environment:

(1) For *acidic* media: In an aqueous environment of strong acid (e.g., HNO_3, HCl, H_2SO_4), add H^+ in requisite numbers to the side deficient therein. The maintenance in the equation of this species freely and independently is completely consistent with an acidic environment.

In an aqueous environment of weak acid (e.g., $HC_2H_3O_2$, H_2CO_3, H_2S), introduction into the equation of requisite H^+ remains perfectly consistent with its existence in solution as such. Nonetheless, a logical appraisal of the total chemistry being portrayed by the equation would (considering the emphasis placed by the ion-electron method upon the chemical compatibilities of species and the realities of their specific identities) invite inclusion of the ionization of the weak acid as a mechanism of progressively supplied H^+ ion, and which, in itself, represents a chemical process.

For this reason, it would be justifiable to meet the requirements for H^+ ions in media of weakly ionized acids by introducing into the equation on the side deficient in H atoms (assuming that the weak acid is involved solely in monoprotic ionization) the molecular species (or polyatomic parent species if ionic) of the weak acid in numbers equal to the hydrogen atoms that are needed. The opposite side of the equation should then be counterbalanced with the conjugate residual species in numbers equal to the molecular species introduced in making up the deficiency of the hydrogen atoms. Thus, in a medium of aqueous acetic acid ($HC_2H_3O_2$), a deficiency of one H atom may be compensated for by expressing the requisite arithmetic in logical compatibility with the following requirements of mass conservation,

$$[\textit{net deficiency of } 1\text{H } \textit{atom}] + \text{HC}_2\text{H}_3\text{O}_2 \rightarrow \text{C}_2\text{H}_3\text{O}_2^-;$$

or

$$\text{C}_2\text{H}_3\text{O}_2^- \rightarrow [\textit{net deficiency of } 1\text{H } \textit{atom}] + \text{HC}_2\text{H}_3\text{O}_2$$

$$[\textit{net deficiency of } x\text{H } \textit{atoms}] + x\text{HC}_2\text{H}_3\text{O}_2 \rightarrow x\text{C}_2\text{H}_3\text{O}_2^-;$$

or

$$x\text{C}_2\text{H}_3\text{O}_2^- \rightarrow [\textit{net deficiency of } x\text{H } \textit{atoms}] + x\text{HC}_2\text{H}_3\text{O}_2.$$

(2) For *alkali-basic* media: In NaOH or KOH environment add H_2O in requisite numbers to the side deficient in hydrogen atoms, using one H_2O molecule for each H atom required and counterbalancing on the opposite side with OH^- ions in numbers equal to that of the H_2O molecules that have been introduced.

The validity of this procedure is manifest in visualizing H_2O as H—OH, and in the arithmetical correctness of the *mass* representations

$$\text{OH}^- \rightarrow [\textit{net deficiency of } 1\text{H } \textit{atom}] + \text{H}_2\text{O};$$

or

$$[\textit{net deficiency of } 1\text{H } \textit{atom}] + \text{H}_2\text{O} \rightarrow \text{OH}^-$$

$$x\text{OH}^- \rightarrow [\text{net deficiency of } x\text{H } atoms] + x\text{H}_2\text{O};$$

or

$$[\textit{net deficiency of } x\text{H } atoms] + x\text{H}_2\text{O} \rightarrow x\text{OH}^-.$$

This procedure is necessary in order to avoid the prohibited introduction and maintenance of significantly expressed H^+ in a reaction that is rich in OH^- ion.

(3) For *ammonia-basic (ammoniacal)* media: In the chemical environment of this weak base, add to the side deficient in hydrogen atoms one NH_4^+ ion for each H required, counterbalancing thereafter with NH_3 molecules on the other side in numbers equal to the NH_4^+ ions that have thus been introduced. This valid conformation of the NH_4^+ ion to the representation as H—NH_3, and the consequent justification of the following requirements of mass conservation,

$$\text{NH}_3 \rightarrow [\textit{net deficiency of } 1\text{H } \textit{atom}] + \text{NH}_4^+$$

or

$$\text{NH}_4^+ + [\textit{net deficiency of } 1\text{H } \textit{atom}] \rightarrow \text{NH}_3$$

$$x\text{NH}_3 \rightarrow [\textit{net deficiency of } x\text{H } \textit{atoms}] + x\text{NH}_4^+;$$

or

$$x\mathrm{NH_4^+} + [\textit{net deficiency of }x\mathrm{H}\textit{ atoms}] \rightarrow x\mathrm{NH_3}$$

protects against the manifest inconsistency of maintaining any significant unneutralized $\mathrm{H^+}$ in a solution rich in molecular $\mathrm{NH_3}$.

It might seem perfectly compatible with the given environment of the reaction to supply needed H atoms by introducing highly abundant $\mathrm{H_2O}$ molecules, and then compensate by balancing $\mathrm{OH^-}$ ions on the opposite side. Concentration is an important factor in translating the potentiality of a species to function chemically to that of its active capacity to do so. The extremely meager quantity of $\mathrm{OH^-}$ in comparison with molecular $\mathrm{NH_3}$ would suggest the initial prominence of $\mathrm{NH_3}$ as the active neutralizer rather than $\mathrm{OH^-}$ ion, despite the fact that the latter, in equivalent quantity, is most certainly the stronger base. This has been pointed out in chapter 2.

3. *Balance equation electronically.*

Having now dutifully complied in all respects with the law of conservation of mass, the equation is ready to fulfill the law of conservation of electrical energy.

(a) Algebraically total the ionic charges on each side of the equation, and insert the requisite numbers of electrons on the side requiring such additional negative charge, either by direct increase in compensation of a net deficiency, or by diminution of excess positive charge.

If the half-reaction being balanced is one of oxidation, the added electrons appear on the right-hand side, thus representing an energy-product of the reaction and algebraically reading plus $(+)$ electrons. If the partial is one of reduction, the added electrons appear on the left-hand side, again algebraically reading plus $(+)$ electrons, and they connote an energy-reactant. The oxidation and reduction phases of the reaction then reveal themselves individually in the two separate half-reactions.

(b) Equalize the preceding electron-energy values by multiplying each partial by the smallest whole number that will produce a net equal transference (loss) of electrons from the reductant with net equal acceptance (gain) of electrons by the oxidant.

Then add the two partials and with the resultant cancellation of the equal numbers of electrons on both sides of the equation, we obtain the final over-all net equation of ionic redox reaction. The ionic equation may be transposed to molecular stoichiometry, if desired, but the ionic equation tells the full and precise story of reaction in conformity with the simple procedure already described under balancing by oxidation numbers.

ILLUSTRATIONS OF ION–ELECTRON BALANCING IN ALL MEDIA

Expositions herein are provided for all three types of aqueous reaction environments that have been described: *acidic, alkali-base* (alkaline), and *ammonia-basic* (ammoniacal). The stepwise balancing procedure perforce, is somewhat extended for clarity of descriptive portrayal, but when familiarity with correct procedure is acquired — by careful pursuance of the systematic progressive sequences herein — the equation-balancer is able, quickly and with sureness, to project all necessary thoughts directly into the once-written "partial" without repetition or re-statement in any other form.

For the student who would seek a better appreciation of the comparative merits of the methods of balancing by oxidation number and by ion–electron in point of view of time, of physical effort, and of a comprehension of the actual chemistry of the species — it is recommended that the equations previously offered as examples in balancing by the section on oxidation numbers be rebalanced by the ion–electron method. It is recommended that the following samples of ion–electron balancing be re-performed, using the oxidation-number method.

Ion–Electron Redox in Aqueous Acidic Medium

■ EXAMPLE 1.   *Strong acidic medium.*

Dichromate ion + thiosulfate ion + dilute sulfuric acid   to yield
chromic ion + sulfate ion.

(a) Set up *skeletal equation* for *first* partial (either one):
$$Cr_2O_7^{2-} \rightarrow Cr^{3+}.$$

(b) Balance equation *atomically:*

Atoms other than O or H: $Cr_2O_7^{2-} \rightarrow 2Cr^{3+}$
Oxygen atoms: $Cr_2O_7^{2-} \rightarrow 2Cr^{3+} + 7H_2O$
Hydrogen atoms: $Cr_2O_7^{2-} + 14H^+ \rightarrow 2Cr^{3+} + 7H_2O.$

(c) Balance equation *electronically*:
$$Cr_2O_7^{2-} + 14H^+ + 6e^- \rightarrow 2Cr^{3+} + 7H_2O$$

(d) Set up *skeletal equation* for *second* partial:
$$S_2O_3^{2-} \rightarrow SO_4^{2-}.$$

(e) Balance equation *atomically:*

Atoms other than O or H: $S_2O_3^{2-} \rightarrow 2SO_4^{2-}$
Oxygen atoms: $S_2O_3^{2-} + 5H_2O \rightarrow 2SO_4^{2-}$
Hydrogen atoms: $S_2O_3^{2-} + 5H_2O \rightarrow 2SO_4^{2-} + 10H^+.$

(f) Balance equation *electronically:*

$$S_2O_3^{2-} + 5H_2O \rightarrow 2SO_4^{2-} + 10H^+ + 8e^-.$$

With each of the respective partials individually complete,

(g) *Equalize electron transfer* to obtain net equation of over-all redox:

$$4 \times [Cr_2O_7^{2-} + 14H^+ + 6e^- \rightarrow 2Cr^{3+} + 7H_2O]$$
$$3 \times [S_2O_3^{2-} + 5H_2O \rightarrow 2SO_4^{2-} + 10H^+ + 8e^-]$$

Additive net: $4Cr_2O_7^{2-} + 3S_2O_3^{2-} + 56H^+ + 15H_2O$
$$\rightarrow 8Cr^{3+} + 6SO_4^{2-} + 30H^+ + 28H_2O$$
Identical species collated: $4Cr_2O_7^{2-} + 3S_2O_3^{2-} + 26H^+$
$$\rightarrow 8Cr^{3+} + 6SO_4^{2-} + 13H_2O.$$

■ EXAMPLE 2: *Weak acidic medium.*

Cobaltous ion + potassium nitrite in acetic acid   to yield
potassium hexanitrocobaltate(III) and nitric oxide.

(a) Set up *skeletal equation* for *first* partial:

$$Co^{2+} \rightarrow K_3Co(NO_2)_6.$$

(b) Balance equation *atomically:*

Atoms other than O or H: $Co^{2+} + 3K^+ + 6NO_2^- \rightarrow K_3Co(NO_2)_6.$
Oxygen atoms: None additional required.
Hydrogen atoms: None required.

(c) Balance equation *electronically:*

$$Co^{2+} + 3K^+ + 6NO_2^- \rightarrow K_3Co(NO_2)_6 + 1e^-.$$

(d) Set up *skeletal equation* for *second* partial:

$$NO_2^- \rightarrow NO$$

(e) Balance equation *atomically:*

With respect to atoms other than O or H: $NO_2^- \rightarrow NO.$
With respect to oxygen atoms: $NO_2^- \rightarrow NO + H_2O$
With respect to hydrogen atoms: $NO_2^- + 2HC_2H_3O_2 \rightarrow NO.$
$$+ H_2O + 2C_2H_3O_2^-.$$

(f) Balance *electronically:*

$$NO_2^- + 2HC_2H_3O_2 + 1e^- \rightarrow NO + H_2O + 2C_2H_3O_2^-.$$

With each of the respective partials individually complete,

(g) *Equalize electron transfer* to obtain net equation of over-all redox:

$$1 \times [Co^{2+} + 3K^+ + 6NO_2^- \rightarrow K_3Co(NO_2)_6 + 1e^-]$$
$$1 \times [NO_2^- + 2HC_2H_3O_2 + 1e^- \rightarrow NO + H_2O + 2C_2H_3O_2^-]$$

Additive net: $Co^{2+} + 3K^+ + 7NO_2^- + 2HC_2H_3O_2 \rightarrow K_3Co(NO_2)_6$
$$+ NO + H_2O + 2C_2H_3O_2^-.$$

■ EXAMPLE 3:   *Multiple redox in acidic medium.*

Arsenious sulfide + nitric acid   to yield

orthoarsenate ion and elemental sulfur

(a) Set up *skeletal equation* for *first* partial:

$$As_2S_3 \rightarrow H_2AsO_4^- + S^0.$$

(b) Balance equation *atomically:*

Atoms other than O or H: $As_2S_3 \rightarrow 2H_2AsO_4^- + 3S^0$
        Oxygen atoms: $As_2S_3 + 8H_2O \rightarrow 2H_2AsO_4 + 3S^0$
        Hydrogen atoms: $As_2S_3 + 8H_2O \rightarrow 2H_2AsO_4 + 3S^0 + 12H^+.$

(c) Balance equation *electronically:*

$$As_2S_3 + 8H_2O \rightarrow 2H_2AsO_4^- + 3S^0 + 12H^+ + 10e^-.$$

(d) Set up *skeletal equation* for *second* partial:

$$NO_3^- \rightarrow NO.$$

(e) Balance equation *atomically:*

    Atoms other than O or H: $NO_3^- \rightarrow NO$
        Oxygen atoms: $NO_3^- \rightarrow NO + 2H_2O$
        Hydrogen atoms: $NO_3^- + 4H^+ \rightarrow NO + 2H_2O.$

(*f*) Balance equation *electronically:*

$$NO_3^- + 4H^+ + 3e^- \rightarrow NO + 2H_2O$$

With each of the respective partials individually complete with them

(g) *Equalize electron transfer* to obtain net equation of over-all redox:

$$3 \times [As_2S_3 + 8H_2O \rightarrow 2H_2AsO_4^- + 3S^0 + 12H^+ + 10e^-]$$
$$10 \times [NO_3^- + 4H^+ + 3e^- \rightarrow NO + 2H_2O]$$

Additive net: $3As_2S_3 + 10NO_3^- + 24H_2O + 40H^+ \rightarrow 6H_2AsO_4^-$
$$+ 10NO + 9S^0 + 20H_2O + 36H^+$$
Identical species collated: $3As_2S_3 + 10NO_3^- + 4H_2O + 4H^+$
$$\rightarrow 6H_2AsO_4^- + 10NO + 9S^0.$$

■ EXAMPLE 4: *Disproportionation in acidic medium.*

Ferric sulfide + hydrochloric acid   to yield
$$\text{ferrous ion and hydrogen sulfide}$$

(a) Set up the *skeletal equation* for *first* partial:

$$Fe_2S_3 \rightarrow Fe^{2+} + S^0.$$

(b) Balance equation *atomically:*

Atoms other than O *or* H: $Fe_2S_3 \rightarrow 2Fe^{2+} + 3S^0.$
Oxygen atoms: *None* required.
Hydrogen atoms: *None* required.

(c) Balance equation *electronically:*

$$Fe_2S_3 \rightarrow 2Fe^{2+} + 3S^0 + 4e^-$$

(d) Set up *skeletal equation* for *second* partial:

$$Fe_2S_3 \rightarrow Fe^{2+} + H_2S.$$

(e) Balance equation *atomically:*

Atoms other than O *or* H: $Fe_2S_3 \rightarrow 2Fe^{2+} + 3H_2S.$
Oxygen atoms: None required.
Hydrogen atoms: $Fe_2S_3 + 6H^+ \rightarrow 2Fe^{2+} + 3H_2S.$

(f) Balance equation *electronically:*

$$Fe_2S_3 + 6H^+ + 2e^- \rightarrow 2Fe^{2+} + 3H_2S.$$

With each of the respective partials individually complete,

(g) *Equalize electron transfer* to obtain net equation of over-all redox:

$$1 \times [Fe_2S_3 \rightarrow 2Fe^{2+} + 3S^0 + 4e^-]$$
$$2 \times [Fe_2S_3 + 6H^+ + 2e^- \rightarrow 2Fe^{2+} + 3H_2S]$$

Additive net: $3Fe_2S_3 + 12H^+ \rightarrow 6Fe^{2+} + 3S^0 + 6H_2S.$

Ion–Electron Redox in Aqueous Alkali-Basic Medium

■ EXAMPLE 1: *Strong alkali-basic medium.*

Chlorine + chromic iodide + sodium hydroxide   to yield
$$\text{chloride ion + chromate ion + (meta) periodate ion}$$

(a) Set up *skeletal equation* for *first* partial:

$$Cl_2 \rightarrow Cl^-.$$

(b) Balance equation *atomically:*

$$\begin{aligned}
\textit{Atoms other than } O \textit{ or } H: & \quad Cl_2 \rightarrow 2Cl^- \\
\textit{Oxygen atoms:} & \quad \textit{none} \text{ required.} \\
\textit{Hydrogen atoms:} & \quad \textit{none} \text{ required.}
\end{aligned}$$

(c) Balance equation *electronically:*

$$Cl_2 + 2e^- \rightarrow 2Cl^-.$$

(d) Set up *skeletal equation* for *second* partial:

$$CrI_3 \rightarrow CrO_4^{2-} + IO_4^-.$$

(e) Balance equation *atomically:*

$$\begin{aligned}
\textit{Atoms other than } O \textit{ or } H: & \quad CrI_3 \rightarrow CrO_4^{2-} + 3IO_4^- \\
\textit{Oxygen atoms:} & \quad CrI_3 + 16OH^- \rightarrow CrO_4^{2-} + 3IO_4^- \\
\textit{Hydrogen atoms:} & \quad CrI_3 + \underbrace{16OH^- + 16OH^-}_{= 32OH^-} \\
& \quad \rightarrow CrO_4^{2-} + 3IO_4^- + 16H_2O.
\end{aligned}$$

(f) Balance equation *electronically:*

$$CrI_3 + 32OH^- \rightarrow CrO_4^{2-} + 3IO_4^- + 16H_2O + 27e^-.$$

With *each* of the respective partials individually complete,

(g) *Equalize electron transfer* to obtain net equation of over-all redox:

$$\begin{array}{l}
27 \times [Cl_2 + 2e^- \rightarrow 2Cl^-] \\
2 \times [CrI_3 + 32OH^- \rightarrow CrO_4^{2-} + 3IO_4^- + 16H_2O + 27e^-] \\
\hline
\end{array}$$

Additive net: $27Cl_2 + 2CrI_3 + 64OH^- \rightarrow 54Cl^-$
$$+ 2CrO_4^{2-} + 6IO_4^- + 32H_2O.$$

■ EXAMPLE 2:   *Disproportionation in strong alkali-basic medium.*

Bromine + hot potassium hydroxide   to yield

bromate ion and bromide ion

(a) Set up *skeletal equation* for *first* partial:

$$Br_2 \rightarrow Br^-$$

(b) Balance equation *atomically:*

$$\begin{aligned}
\textit{Atoms other than } O \textit{ or } H: & \quad Br_2 \rightarrow 2Br^- \\
\textit{Oxygen atoms:} & \quad \textit{none} \text{ required.} \\
\textit{Hydrogen atoms:} & \quad \textit{none} \text{ required.}
\end{aligned}$$

(c) Balance equation *electronically:*

$$Br_2 + 2e^- \rightarrow 2Br^-.$$

(d) Set up *skeletal equation* for *second* partial:
$$Br_2 \rightarrow BrO_3^-.$$

(e) Balance equation *atomically:*

Atoms other than O or H: $Br_2 \rightarrow 2BrO_3^-.$

Oxygen atoms: $Br_2 + 6OH^- \rightarrow 2BrO_3^-.$

Hydrogen atoms: $Br_2 + \underbrace{6OH^- + 6OH^-}_{= 12OH^-} \rightarrow 2BrO_3^- + 6H_2O.$

(f) Balance equation *electronically:*
$$Br_2 + 12OH^- \rightarrow 2BrO_3^- + 6H_2O + 10e^-.$$

With each of the respective partials individually complete,

(g) *Equalize electron transfer* to obtain net equation of over-all redox:

$$5 \times [Br_2 + 2e^- \rightarrow 2Br^-]$$
$$1 \times [Br_2 + 12OH^- \rightarrow 2BrO_3^- + 6H_2O + 10e^-]$$

Additive: $6Br_2 + 12OH^- \rightarrow 10Br^- + 2BrO_3^- + 6H_2O.$

Simplification: $3Br_2 + 6OH^- \rightarrow 5Br^- + BrO_3^- + 3H_2O.$

Ion–Electron Redox in Ammonia-Basic Medium

■ EXAMPLE 1:   *Ammoniacal medium.*

Cupric ammonia complex ion + excess sodium cyanide in excess
ammonia   to yield   cuprous cyanide complex ion and cyanate ion

(a) Set up *skeletal equation* for *first* partial:
$$Cu(NH_3)_4^{2+} \rightarrow Cu(CN)_3^{2-}$$

(b) Balance equation *atomically:*

Atoms other than O or H: $Cu(NH_3)_4^{2+} + 3CN^- \rightarrow Cu(CN)_3^{2-} + 4NH_3$

Oxygen atoms: *none* required

Hydrogen atoms: *none* additional required

(c) Balance equation *electronically:*
$$Cu(NH_3)_4^{2+} + 3CN^- + 1e^- \rightarrow Cu(CN)_3^{2-} + 4NH_3$$

(d) Set up *skeletal equation* for *second* partial:
$$CN^- \rightarrow CNO^-$$

(e) Balance equation *atomically:*

Atoms other than O or H: $CN^- \rightarrow CNO^-$

Oxygen atoms: $CN^- + H_2O \rightarrow CNO^-$

Hydrogen atoms: $CN^- + H_2O + 2NH_3 \rightarrow CNO^- + 2NH_4^+$

(f) Balance equation *electronically:*

$$CN^- + H_2O + 2NH_3 \rightarrow CNO^- + 2NH_4^+ + 2e^-$$

With *each* of the respective partials individually complete,

(g) *Equalize electron transfer* to obtain net equation of over-all redox:

$$2 \times [Cu(NH_3)_4^{2+} + 3CN^- + 1e^- \rightarrow Cu(CN)_3^{2-} + 4NH_3]$$
$$1 \times [CN^- + H_2O + 2NH_3 \rightarrow CNO^- + 2NH_4^+ + 2e^-]$$

Additive net: $2Cu(NH_3)_4^{2+} + 7CN^- + H_2O + 2NH_3$
$$\rightarrow 2Cu(CN)_3^{2-} + CNO^- + 2NH_4^+ + 8NH_3$$
Identical species collated: $2Cu(NH_3)_4^{2+} + 7CN^- + H_2O$
$$\rightarrow 2Cu(CN)_3^{2-} + CNO^- + 2NH_4^+ + 6NH_3.$$

■ EXAMPLE 2: *Disproportionation in ammoniacal environment.*

Mercurous chloride + aqueous ammonia to yield mercuric amido chloride and metallic mercury.

(a) Set up *skeletal equation* for *first* partial:

$$Hg_2Cl_2 \rightarrow Hg(NH_2)Cl.$$

(b) Balance equation *atomically:*

Atoms other than O or H: $\quad Hg_2Cl_2 + 2NH_3 \rightarrow 2Hg(NH_2)Cl.$
Oxygen atoms: $\quad$ *none* required.
Hydrogen atoms: $\quad Hg_2Cl_2 + \underbrace{2NH_3 + 2NH_3}_{4NH_3} \rightarrow 2Hg(NH_2)Cl. + 2NH_4^+$

(c) Balance equation *electronically:*

$$Hg_2Cl_2 + 4NH_3 \rightarrow 2Hg(NH_2)Cl + 2NH_4^+ + 2e^-$$

(d) Set up the *skeletal equation* for *second* partial:

$$Hg_2Cl_2 \rightarrow Hg^0$$

(e) Balance equation *atomically:*

Atoms other than O or H: $\quad Hg_2Cl_2 \rightarrow 2Hg^0 + 2Cl^-$
Oxygen atoms: $\quad$ *none* required
Hydrogen atoms: $\quad$ *none* required

(f) Balance equation *electronically:*

$$Hg_2Cl_2 + 2e^- \rightarrow 2Hg^0 + 2Cl^-$$

With *each* of the respective partials individually complete, we proceed with them to

(g) *Equalize electron transfer* to obtain the net equation of over-all redox

$$1 \times [Hg_2Cl_2 + 4NH_3 \rightarrow 2Hg(NH_2)Cl + 2NH_4^+ + 2e^-]$$
$$1 \times [Hg_2Cl_2 + 2e^- \rightarrow 2Hg^0 + 2Cl^-]$$

Additive net: $2Hg_2Cl_2 + 4NH_3 \rightarrow 2Hg(NH_2)Cl + 2Hg^0 + 2Cl^- + 2NH_4^+$
Simplification: $Hg_2Cl_2 + 2NH_3 \rightarrow Hg(NH_2)Cl + Hg^0 + Cl^- + NH_4^+$.

Exercises

1. The chemical products of reaction in each of the following skeletal equations are fully stated. Balance each by oxidation numbers and check the accuracy of the result in Appendix A (ANSWERS).

   (a) $Hg_2CrO_4 \rightarrow Hg^0 + Cr_2O_3 + O_2$
   (b) $Co^{2+} + O_2 + NH_3 + H_2O \rightarrow Co(NH_3)_6^{3+} + OH^-$
   (c) $K_2Cr_2O_7 + NaCl + H_2SO_4 \rightarrow CrO_2Cl_2 + KHSO_4 + NaHSO_4 + H_2O$
   (d) $NaHSO_3 + NaIO_3 \rightarrow NaHSO_4 + Na_2SO_4 + I_2 + H_2O$
   (e) $Cr_2O_3 + K_2CO_3 + O_2 \rightarrow K_2CrO_4 + CO_2$
   (f) $Au + KCN + O_2 + H_2O \rightarrow KAu(CN)_2 + KOH$
   (g) $CH_3OH + NaClO_3 + H_2SO_4 \rightarrow CO_2 + ClO_2 + Na_2SO_4 + H_2O$
   (h) $HI + H_2SO_4 \rightarrow I_2 + H_2S + H_2O$
   (i) $Ca_3(PO_4)_2 + SiO_2 + C \rightarrow CaSiO_3 + P_4 + CO$
   (j) $KClO_3 + HCl \rightarrow KCl + ClO_2 + Cl_2 + H_2O$
   (k) $Fe(CrO_2)_2 + Na_2CO_3 + O_2 \rightarrow Fe_2O_3 + Na_2CrO_4 + CO_2$
   (l) $MnO_2 + Na_2CO_3 + KClO_3 \rightarrow Na_2MnO_4 + KCl + CO_2$
   (m) $Fe(CN)_6^{4-} + SO_4^{2-} + H_2O + H^+ \rightarrow Fe^{2+} + HSO_4^- + NH_4^+ + CO$
   (n) $TiCl_4 + Na_2S_2O_3 + H_2O \rightarrow H_2TiO_3 + NaCl + SO_2 + S^0$
   (o) $(NH_4)_2SO_4 \rightarrow N_2 + NH_3 + SO_2 + H_2O$.

2. Provide complete and balanced ionic equations for the following reactions in the aqueous media stated. Then convert to molecular stoichiometry and check the accuracy of the result in Appendix A (ANSWERS).

   (a) $P_4 \xrightarrow{KOH} PH_3 + H_2PO_2^-$
   (b) $[K^+ + MnO_4^-] \xrightarrow{HCl} Mn^{2+} + Cl_2$
   (c) $As_2O_3 + Zn^0 \xrightarrow{HCl} As^0 + Zn^{2+} + H_2$
   (d) $Al^0 + [Na^+ + H_2AsO_3^-] \xrightarrow{NaOH} Al(OH)_4^- + AsH_3$
   (e) $CdS \xrightarrow{HNO_3} Cd^{2+} + S^0 + NO$
   (f) $[Fe^{2+} + SO_4^{2-}] + [Na^+ + O_2^{2-}] \xrightarrow{NaOH} Fe(OH)_3 + OH^-$
   (g) $F_2 + OH^- \xrightarrow{NaOH} OF_2$
   (h) $PbO_2 + [Mn^{2+} + NO_3^-] \xrightarrow{HNO_3} Pb^{2+} + MnO_4^-$
   (i) $U_3O_8 \xrightarrow[\text{aqua regia}]{HCl + HNO_3} (UO_2)Cl_2 + NO$
   (j) $FeS \xrightarrow{HNO_3} Fe^{2+} + S^0 + NO$
   (k) $[Cu(NH_3)_4^{2+} + SO_4^{2-}] + [Na^+ + S_2O_4^{2-}] \xrightarrow{NH_3} Cu^0 + SO_3^{2-}$
   (l) $[Ti^{3+} + SO_4^{2-}] + [K^+ + MnO_4^-] \xrightarrow{H_2SO_4} Ti^{4+} + Mn^{2+}$
   (m) $[Mn^{2+} + SO_4^{2-}] + O_2 \xrightarrow{NH_3} Mn(OH)_3 + H_2O$.

3. All the following reactions occur, under appropriate conditions of temperature and concentration, in the aqueous acidic or basic media of the chemical character indicated. Complete and balance each by the ion-electron method of half-reactions and check the accuracy of the resultant net equation of over-all redox within Appendix A (ANSWERS).

(a) $Co(NO_2)_6^{3-} + NH_4^+ \xrightarrow{HC_2H_3O_2} Co^{2+} + N_2$

(b) $Fe(OH)_3 + Cl_2 \xrightarrow{KOH} FeO_4^{2-} + Cl^-$

(c) $BiO(OH) + HSnO_4^- \xrightarrow{NaOH} Bi^0 + Sn(OH)_6^{2-}$

(d) $C_2H_5OH + Cr_2O_7^{2-} \xrightarrow{H_2SO_4} HC_2H_3O_2 + Cr^{3+}$

(e) $SbO^+ + S_2O_3^{2-} \xrightarrow{HC_2H_3O_2} Sb_2OS_2 + SO_2$

(f) $P_4 \xrightarrow{KOH} PH_3 + H_2PO_2^-$

(g) $IO_3^- + I_2 \xrightarrow{HCl} ICl$

(h) $CrO_5 \xrightarrow{HNO_3} Cr^{3+} + O_2$

(i) $S_2O_8^{2-} + Mn^{2+} \xrightarrow{HNO_3} MnO_4^- + SO_4^{2-}$

(j) $Fe^{3+} + C_2H_3OS^- \xrightarrow{NH_3} FeS + C_2H_3O_2^- + S^0$

(k) $As_2S_3 + H_2O_2 \xrightarrow{NH_3} AsO_4^{3-} + SO_4^{2-} + H_2O$

(l) $Au^0 \xrightarrow[\text{aqua regia}]{HCl + HNO_3} AuCl_4^- + NO_2$

(m) $Fe(SCN)^{2+} + MnO_4^- \xrightarrow{H_2SO_4} Fe^{3+} + SO_2 + CO_2 + NO_3^- + Mn^{2+}$

(n) $Mn^{2+} + BiO_3^- \xrightarrow{HNO_3} MnO_4^- + Bi^{3+}$

(o) $Hg_2^{2+} + NO_3^- \xrightarrow{NH_3} Hg_2O(NH_2)NO_3 + Hg^0 \ldots$ [hint: (NH$_2$) from NH$_3$]

(p) $Al^0 + NO_3^- \xrightarrow{NaOH} Al(OH)_4^- + NH_3$

(q) $Fe(C_6H_5O_7)_2^{3-} + H_2S \xrightarrow{NH_3} FeS + S_2^{2-}$

(r) $Sb_2O_3 + Ag^+ \xrightarrow{NH_3} Sb_2O_5 + Ag^0$

(s) $PtCl_6^{2-} + C_3H_5(OH)_3 \xrightarrow{NaOH} CO_3^{2-} + C_2O_4^{2-} + Pt^0$

(t) $SCN^- + NO_3^- \xrightarrow{HNO_3} SO_4^{2-} + CO_2 + NO$

(u) $Pb^0 + C_2H_3O_2^- + O_2 \xrightarrow{HC_2H_3O_2} Pb(C_2H_3O_2)_4^{2-} + H_2O$

(v) $HgS \xrightarrow[\text{aqua regia}]{HCl + HNO_3} HgCl_4^{2-} + S^0 + NO.$

CHAPTER 8

CHEMICAL EQUIVALENCE AND VOLUMETRIC STOICHIOMETRY

The conceptual theme of the presentation in this chapter is *chemical equivalence*. Although definitions have been made separately elsewhere, the importance of this concept in volumetric stoichiometry more than justifies any repetitions that may be encountered here as part of an organized, orderly development. Whatever may be the mechanisms of chemical reactions or the infinite variability in conditions under which they occur, their quantitative aspects are defined by the equality of electrons lost and gained, respectively, in the mutual interactions of reductants and oxidants. In metathetical reaction (non-redox), they are defined by the equality of the replacements, returns, exchanges, or associations of positive and negative charges without permanent gain or loss to the reaction species. All this is just another way of saying that reaction always occurs "equivalent for equivalent."

The very device that permits of an orderly and systematic balancing of redox equations, that keeping track of shifts of electrons and thus ensuring that the laws of conservation of energy and of mass are fulfilled, underlies the systematic arithmetical approaches to the stoichiometry of reactions. We delineate separately the methods of determining equivalent weights in each of the two broad categories of reactants, metathetical and redox, bearing in mind that the same substance may portray roles in both categories.

In either instance, we are concerned with the Avogadro constant, 6.02×10^{23}.

DETERMINING METATHETICAL EQUIVALENTS

How does one deduce the metathetical (nonredox) gram-equivalent weight of an acid, base, or salt? Inherent in this evaluation must be the interpretation of "metathesis" as merely an exchange of positively or

167

negatively charged partners; or of their mutual sharing of electrons without undergoing any permanent loss or gain in oxidation state or of valence.

Let us consider nitric acid, HNO_3, as an acid. If this added phrase seems at first glance quite redundant, remember that HNO_3 may also act as an oxidant — in which event, we would concern ourselves with its NO_3^- ion content, rather than with its H^+ ion supply. Clearly, a mole of HNO_3, a monoprotic acid, can deliver only one mole of H^+ ion for a gram-formula weight of the parent solute in response to the neutralizing demands of a base; that is, the acid can transport an associated Avogadro number of positive charges (H^+ ions) for a gram-formula weight. Hence, in accordance with definition, the required weight of HNO_3 is

$$\frac{63.0 \text{ g/mole (g-formula)}}{1 \text{ g-equiv/mole (g-formula)}} = 63.0 \text{ g to g-equiv.}$$

The sizes of the units might have been reduced in conformity with the relationships

$$\frac{63.0 \text{ mg/mmole (mg-formula)}}{1 \text{ mg-equiv (mequiv)/mmole (mg-formula)}}$$

$$= 63.0 \text{ mg to a milliequivalent.}$$

As observed, the numerical value remains unchanged. The "milli-" scale, obviously, proves of distinct convenience in contending theoretically and experimentally with the generally small volumes and weights of academic practice.

Let us now turn to a polyprotic acid, H_3PO_4. To assume, in the absence of further information, that the gram-equivalent weight of this acid would be one-third of its gram-formula weight solely because one whole mole of it, in itself, potentially provides three times the number of H^+ ions necessary to constitute the requisite Avogadro number of positive charges, might well, indeed, prove to be a highly inaccurate guess. Although a polyprotic acid can supply more than one H^+ ion for a formula unit of the parent acid, the delivery depends upon the demands made upon it by the variable quantity of base that may be present. Complete neutralization of the acid, if stipulated, would require the delivery of three H^+ ions to the base but, just as obvious must be the partial neutralizations that occur when the quantities of base are limited. Thus, three different gram-equivalent weights are then possible for phosphoric acid, H_3PO_4, conforming, respectively, to the following reactions:

(a) $H_3PO_4 + OH^- \rightarrow H_2O + H_2PO_4^-$ *one* mole of H^+,

(b) $H_3PO_4 + 2OH^- \rightarrow 2H_2O + HPO_4^{2-}$ *two* moles of H^+,

(c) $H_3PO_4 + 3OH^- \rightarrow 3H_2O + PO_4^{3-}$ *three* moles of H^+
                                                            in formula weight of
                                                            acid neutralized.

It must be emphasized that the gram-equivalent weight of a substance conforms to its *actual* behavior in reaction rather than to its potential behavior. Each of three different weights of H_3PO_4 can transfer, in accordance with definition, one mole of H^+ ion for each gram-formula weight of the acid, but each of these can be correct only for its stated specific reaction. For the preceding reactions, these are:

(a)

$$\frac{98 \text{ g/mole}}{1 \text{ g-equiv/mole}} = 98.0 \text{ of } H_3PO_4 \text{ to a gram-equivalent}$$

(b)

$$\frac{98 \text{ g/mole}}{2 \text{ g-equiv/mole}} = 98.0 \text{ g of } H_3PO_4 \text{ to a gram-equivalent}$$

(c)

$$\frac{98.0 \text{ g/mole}}{3 \text{ g-equiv/mole}} = 32.7 \text{ g of } H_3PO_4 \text{ to a gram-equivalent.}$$

What is being stressed here most urgently is that the reaction product must be known before it is possible to proceed with the determination of the gram-equivalent weight. To state that a liter of a particular solution contains one gram-equivalent weight of H_3PO_4 (1 N solution) without accompanying information with respect to the reaction for which it is calculated, not only does not tell whether 98.0, 49.0, or 32.7 g of the solute have been dissolved therein, but also confounds any procedure, (except experimental analysis) to translate its concentration into terms of in terms of molarity ensure, the necessary conveyance of the precise concentration that it provides, for purposes of determining the gram-equivalent weights and of its consequent normality with respect to the reaction in which it is involved. Molarity is fixed and invariable for any given solution, but normality of the same preparation may have variable values depending upon the conditions of reaction under which the reagent functions.

The interpretations of metathetical equivalents of acids, in general, apply in equal degree to those of bases. By definition, the gram-equivalent weight of an acid represents that weight donating upon demand one mole of H^+ ion for a gram-formula weight of the parent solute. Similarly, the gram-equivalent weight of a base represents that weight of it that supplies, likewise in response to demands of neutralization, one mole of OH^- ion for a gram-formula weight of the parent solute.

According to the Brönsted-Lowry concept, a base is that weight that actually accepts one mole of H^+ ion for a gram-formula weight of the base in the reaction of neutralization. Although the gram-equivalent weight of NaOH would (thus for any neutralization) be the invariable quantity

$$\frac{40.0 \text{ g/mole}}{1 \text{ g-equiv/mole}} = 40.0 \text{ g of NaOH to a gram-equivalent,}$$

the gram-equivalent weight of $Ca(OH)_2$ could have two different numerical values, depending upon the specific reaction. Thus, for the *partial* neutralization, expressed by

$$Ca(OH)_2 + H^+ \rightarrow Ca(OH)^+ + H_2O,$$

the gram-equivalent weight would be

$$\frac{74.1 \text{ g/mole}}{1 \text{ g-equiv/mole}} = 74.1 \text{ g of } Ca(OH)_2 \text{ to a gram-equivalent,}$$

whereas, for the *complete* neutralization expressed by

$$Ca(OH)_2 + 2H^+ \rightarrow Ca^{2+} + 2H_2O,$$

the gram-equivalent weight would be

$$\frac{74.1 \text{ g/mole}}{2 \text{ g-equiv/mole}} = 37.1 \text{ g of } Ca(OH)_2 \text{ to a gram-equivalent.}$$

The determination of the gram-equivalent weight of NH_3 is sometimes puzzling to beginners who fail to reconcile the identity of this compound with that depicted by the Arrhenius formulation of NH_4OH, but it responds quite simply to its character as a Brönsted-Lowry proton acceptor,

$$NH_3 + H^+ \rightarrow NH_4^+$$

and, hence, yields

$$\frac{17.9 \text{ g/mole}}{1 \text{ g-equiv/mole}} = 17.0 \text{ g of } NH_3 \text{ to a gram-equivalent.}$$

The metathetical gram-equivalent weight of a *salt*, taken as a gram-formula unit, or of any of its constituent ions, likewise proves to be the molar weight of the species divided by a whole number that converts the total quantity of charge (positive or negative) of the constituent species of the salt to the charge value of the Avogadro constant. This whole number represents the total valence of a mole of the ion of the particular salt to which reference is being made. In a *simple* salt (that is, one containing only a single species of cation and a single species of anion), the total valence of one ion of the salt would have to be, necessarily, identical to that of the other. Hence, any solution containing a gram-equivalent weight of a simple salt must inevitably supply a gram-equivalent weight of each of its constituent ions; that is, a solution that is $1N$ with respect to one must be likewise $1N$ with respect to the other.

In double salts, or complex compounds that can yield more than one kind of cation or anion, the particular constituent for which the gram-equivalent weight is being calculated must be specified, inasmuch as the numerical value of the equivalent weight that is assigned to one need not necessarily coincide with that assigned to another. Hence, in complex or double salts, a given solution that is $1N$ with respect to one constituent ion need not be $1N$ with respect to the others. A few examples will clarify these various aspects —

$$K^+: \frac{39.1 \text{ g/mole or g-ion}}{1 \text{ g-equiv/mole or g-ion}} = \frac{39.1 \text{ g/g-equiv}}{\text{(if dissolved in 1 liter} \rightarrow 1N \text{ K}^+)}$$

$$Cu^{2+}: \frac{63.5 \text{ g/mole or g-ion}}{2 \text{ g-equiv/mole or g-ion}} = \frac{31.8 \text{ g/g-equiv}}{\text{(if dissolved in 1 liter} \rightarrow 1N \text{ Cu}^{2+})}$$

$$Al^{3+}: \frac{27.0 \text{ g/mole or g-ion}}{3 \text{ g-equiv/mole or g-ion}} = \frac{9.0 \text{ g/g-equiv}}{\text{(if dissolved in 1 liter} \rightarrow 1N \text{ Al}^{3+})}$$

$$Cl^-: \frac{35.5 \text{ g/mole or g-ion}}{1 \text{ g-equiv/mole or g-ion}} = \frac{35.5 \text{ g/g-equiv}}{\text{(if dissolved in 1 liter} \rightarrow 1N \text{ Cl}^-)}$$

$$SO_4^{2-}: \frac{96.1 \text{ g/mole or g-ion}}{2 \text{ g-equiv/mole or g-ion}} = \frac{48.1 \text{ g/g-equiv}}{\text{(if dissolved in 1 liter} \rightarrow 1N \text{ SO}_4^{2-})}$$

$$K_2SO_4: \frac{174.3 \text{ g/mole or g-formula}}{2 \text{ g-equiv/mole or g-formula}} = 87.2 \text{ g/equiv}$$
(if dissolved in 1 liter → $1N$ K$^+$ *and* $1N$ SO$_4^{2-}$)

$$KCl: \frac{74.6 \text{ g/mole or g-formula}}{1 \text{ g-equiv/mole or g-formula}} = 74.6 \text{ g/g-equiv}$$
(if dissolved in 1 liter → $1N$ K$^+$ *and* $1N$ Cl$^-$)

$$CuSO_4: \frac{159.6 \text{ g/mole or g-formula}}{2 \text{ g-equiv/mole or g-formula}} = 79.8 \text{ g/g-equiv}$$
(if dissolved in 1 liter → $1N$ Cu^{2+} *and* $1N$ SO$_4^{2-}$)

$$Al_2(SO_4)_3: \frac{342.3 \text{ g/mole or g-formula}}{6 \text{ g-equiv/mole or g-formula}} = 57.1 \text{ g/g-equiv}$$
(if dissolved in a liter → $1N$ Al^{3+} *and* $1N$ SO$_4^{2-}$)

$$KAl(SO_4)_2 . 12H_2O: \frac{474.3 \text{ g/mole or g-formula}}{1 \text{ g-equiv K}^+/\text{mole or g-formula}}$$
(*with respect to* K$^+$)
$$= 474.3 \text{ g/g-equiv}$$
(if dissolved in 1 liter → $1N$ K$^+$, but $3N$ Al^{3+} and $4N$ SO$_4^{2-}$)

(*with respect to* Al^{3+}):
$$\frac{474.3 \text{ g/mole or g-formula}}{3 \text{ g-equiv Al}^{3+}/\text{mole or g-formula}}$$
$$= 158.1 \text{ g/g-equiv}$$
(if dissolved in 1 liter → $1N$ Al^{3+} but $\frac{1}{3}N$ K$^+$ *and* $\frac{4}{3}N$ SO$_4^{2-}$)

(*with respect to* SO$_4^{2-}$):
$$\frac{474.3 \text{ g/mole or g-formula}}{4 \text{ g-equiv SO}_4^{2-}/\text{mole or g-formula}}$$
$$= 118.6 \text{ g/g-equiv}$$
(if dissolved in 1 liter → $1N$ SO$_4^{2-}$, but $\frac{1}{4}N$ K$^+$ and $\frac{3}{4}N$ Al^{3+})

$NaKC_4H_4O_6 \cdot 4H_2O$: (*with respect to* Na^+)

$$\frac{282.2 \text{ g/mole or g-formula}}{1 \text{ g-equiv } Na^+ \text{ or } K^+/\text{mole or g-formula}}$$

$$= 282.2 \text{ g/g-equiv}$$

(if dissolved in 1 liter $\rightarrow 1N \ K^+$ *and* $1N \ Na^+$, *but* $2N \ C_4H_4O_6^{2-}$)

(*with respect to* $C_4H_4O_6^{2-}$):

$$\frac{282. \text{ g/mole or g-formula}}{2 \text{ g-equiv } C_4H_4O_6^{2-}/\text{mole or g-formula}}$$

$$= 141.1 \text{ g/g-equiv}$$

(if dissolved in 1 liter $\rightarrow 1N \ C_4H_4O_6^{2-}$, *but* $\frac{1}{2}N \ K^+$ *and* $\frac{1}{2}N \ Na^+$)

DETERMINING REDOX EQUIVALENTS

We concern ourselves here with the actual loss and gain of electrons that result in *permanent* changes in the initial oxidation states of atoms in interacting species, and/or of the valences of independent ions. The identical considerations apply here as with metathetical equivalents, namely, the derivation from the formula weight (mole) of the species being considered (whether ionic or molecular) of that fractional part (sometimes the entire mole) that involves itself in the transfer of the Avogadro number of electrons. The computation of redox equivalents requires an ever-necessary alertness to the fact that we are concerned therein with the permanent removal or displacement of electrons from one unit of an inter-acting species to another. In essence, we distinguish between valence and change in valence; and between oxidation state or number and change in oxidation state or number. This factor of change represents the critical mathematics that differentiates the determination of redox equivalents from metathetical equivalents (no permanent change).

We have seen that for HNO_3, as an acid, there is present one equivalent to the mole. Let us observe the differences when the same acid is used as an oxidant, that is, for its content of NO_3^- ion. As already emphasized, the products of the specific reaction must be known. We cannot otherwise deduce the changes in electrons involved in the required reduction of the NO_3^- ion if the NO_3^- ion is to function as an oxidant, for without reduction it is impossible to have oxidation, or vice-versa. Let us consider three different reactions of this acid yielding the products NO_2, NO, and NH_4^+ respectively. Ion–electron partials may be written for each to reveal the total change in electrons for a mole of NO_3^- ion; thus

(a) $\qquad NO_3^- + 2H^+ + 1e^- \rightarrow NO_2 + H_2O$,

(b) $\qquad NO_3^- + 4H^+ + 3e^- \rightarrow NO + 2H_2O$,

(c) $\qquad NO_3^- + 10H^+ + 8e^- \rightarrow NH_4^+ + 3H_2O$.

Consequently, *three different equivalent weights* are obtained for the HNO_3 *as an oxidant,* in conformity to the respective reactions so written (others also are possible, involving still-different reaction products). They are, respectively.

(a) $\dfrac{\text{mole HNO}_3}{1} = \begin{array}{l} 1 \ g\text{-}equiv; \\ \text{a mole provides } 1 \end{array}$

(b) $\dfrac{\text{mole HNO}_3}{3} = \begin{array}{l} 1 \ g\text{-}equiv; \\ \text{a mole provides } 3 \end{array}$ equivalents of HNO_3 as an oxidant.

(c) $\dfrac{\text{mole HNO}_3}{8} = \begin{array}{l} 1 \ g\text{-}equiv; \\ 1 \text{ mole provides } 8 \end{array}$

Clearly, to refer to a "$1N$ HNO_3" solution without further qualification imparts not the slightest information as to its mode of employment — as a methathetical *acid* or as an oxidant. Even if the latter is stipulated, there is posed the even greater difficulty of ascertaining (short of analysis) which of the different weights of solute has actually been dissolved in preparing the solution.

Before proceeding further, an earlier observation should be recalled, in the interests of mathematical facility with problems; namely, that it is completely unnecessary to balance an equation solely for the purpose of determining gram-equivalents. This is certainly not intended to convey the impression that chemical equations, which are very necessary in thermo-chemical and other areas of mathematical computation, may be altogether discarded. What is being suggested, rather, is that the concept of gram-equivalents effectively permits escape from the tedium and possibly excessively time-consuming balancing process when incident to stoichiometric calculations of solutions. One may work out weight-mole relationships from the correctly balanced equation in the classical tradition of proportion, but the concept of equivalents when applied to such calculations offers an easier approach. All one needs to know are the initial and final formulas of the species being redoxed, whence the necessary oxidation state changes may be promptly ascertained by appraising the values of the initial and final oxidation state or oxidation number.

Illustratively, the respective gram-equivalent weights of MnO_4^- and Fe^{2+} are readily obtained for their mutual interaction in H_2SO_4 medium upon being informed that the products of redox are Mn^{2+} and Fe^{3+}. As the oxidation state of the manganese atom in MnO_4^- is $+7$ and that of the product is $+2$, clearly 5 electrons have been accepted by the MnO_4^-; hence.

$$\frac{\text{mole MnO}_4^-}{5} = 1 \text{ g-equiv of } MnO_4^- \text{ ion, as an oxidant.}$$

The Fe^{2+}, in being converted to Fe^{3+}, has clearly lost 1 electron; hence,

$$\frac{\text{mole Fe}^{2+}}{1} = 1 \text{ g-equiv of } Fe^{2+} \text{ ion, as a reductant.}$$

Were we to have concerned ourselves with the gram-formulas, $KMnO_4$ and $FeSO_4$, under similar conditions of reactions, with redox products being $MnSO_4$ and $Fe_2(SO_4)_3$, we would have obtained likewise,

$$\frac{\text{mole } KMnO_4}{5} = 1 \text{ g-equiv of } KMnO_4, \text{ as an oxidant.}$$

$$\frac{\text{mole } FeSO_4}{1} = 1 \text{ g-equiv of } FeSO_4, \text{ as a reductant.}$$

If the reaction in acidic medium had been between $Cr_2O_7^{2-}$ and Sn^{2+}, with redox products given as Cr^{3+} and Sn^{4+}, the required gram-equivalent weights of the reactants would then have been evaluated in accordance with the following considerations:

$$\frac{\text{mole } Sn^{2+}}{2} = 1 \text{ g-equiv of } Sn^{2+} \text{ ion, as a reductant, (change in oxidation state of Sn atom from } +2 \text{ to } +4, \text{ an increase of 2 units).}$$

and

$$\frac{\text{mole } Cr_2O_7^{2-}}{6} = 1 \text{ g-equiv of } Cr_2O_7^{2-} \text{ ion, as an oxidant, (change in oxidation state of each Cr atom from } +6 \text{ to } +3 \text{ is a decrease of 3 units, or 6 units for the two Cr atoms in the mole of } Cr_2O_7^{2-}. \text{ The change in oxidation number of the Cr is 6 units.)}$$

As apparent in the preceding example with the $Cr_2O_7^{2-}$ ion, care must always be exercised to ensure that the total change in oxidation number that appears in the denominator conforms with complete consistency to the number of such pertinent atoms designated for the mole (molar weight) as it appears in the numerator.

EQUALITY OF INTERACTING EQUIVALENTS

For any chemically interacting species, A and B, the following relationship must hold for any and all conditions of their interaction:

total gram-equivalents of A = total gram-equivalents of B

or

total milliequivalents of A = total milliequivalents of B.

The computations of electrolyte solutions require, more often than not, the determination of volumes and concentrations. The transposing of gram-equivalents and milliequivalents to the requisite units for purposes of convenient substitution in problems is simple. A $1N$ solution of any substance contains, by definition, one gram-equivalent of that substance to a liter; were two liters of $1N$ solution given, two gram-equivalents would be present. If a $2N$ solution were supplied, one liter of it would provide a total of two gram-equivalents of the solute; and one-half liter of the same solution would contain only one gram-equivalent. Inescapably then,

$$\text{liters} \times \text{normality} = \text{gram-equivalents.}$$

Analogously,

$$\text{milliliters} \times \text{normality} = \text{milliequivalents.}$$

In accordance with the requirements of the problem, we can then interchange at will gram-equivalents for liters-times-normality (and vice-versa); and milliequivalents for milliliters-times-normality (and, likewise, vice-versa). When weight is being employed or sought, the convenient term for substitution will be the *gram-equivalent*, when volume is expressed in *liters;* and *milliequivalent*, when volume is expressed in *milliliters.* When either the volumes or the concentrations of reactant solutions are being employed or sought, the obvious direct expression for evaluation is liters $\times$ normality, or milliliters $\times$ normality. Normality, once obtained, may be converted to other units of concentration as might be required by the problem.

Collating what has been said, we have the relationships

$$\frac{\text{total gram-equivalents}}{\text{liters} \times \text{normality}} = \frac{\text{total gram-equivalents}}{\text{liters} \times \text{normality}}$$

and

$$\frac{\text{total milliequivalents}}{\text{milliliters} \times \text{normality}} = \frac{\text{total milliequivalents}}{\text{milliliters} \times \text{normality.}}$$

The quantitative aspects of these relationships are very clear. Not only do reactions occur *equivalent for equivalent* (or *milliequivalent for milliequivalent*) on a precise basis, but also solutions of the *same normality* inevitably react, if at all, *liter for liter* (or *milliliter for milliliter*).

Let us illustrate the method of operation with a specific example of mathematical application: calculation of the volume, in milliliters, of $0.10M$ $KMnO_4$ solution required to oxidize in acidic medium,

  (a) 30.0 ml of $0.20M$ Fe^{2+},
  (b) 30.0 mg of Fe^{2+},
  (c) 30.0 mmoles of Fe^{2+}.

The products of redox are herewith given as Mn^{2+} and Fe^{3+}.

In each instance, we require for substitution the normality of the $KMnO_4$ solution. As we are given molarity, we must convert to the required unit. The oxidation state of the manganese changes from its initial value of

$+7$ to a final value of $+2$, for a total oxidation number decrease of 5 units. Therefore,

$$\frac{\text{mole } KMnO_4}{5} = 1 \text{ g-equiv of } KMnO_4, \text{ as an oxidant}$$

and, consequently

$$1 \text{ mole of } KMnO_4 = 5 \text{ g-equiv of } KMnO_4, \text{ as an oxidant.}$$

Now, if the mole is contained in one liter, we have a $1M$ solution; hence, with the 5 gram-equivalents necessarily likewise contained in the same liter, the concentration of the latter must also be $5N$. We have thus established the simple equality required for conversion of molarity to normality (and/or vice versa). As an oxidant,

$$1M \; KMnO_4 = 5N \; KMnO_4$$

hence,

$$0.10M \; KMnO_4 = (0.10 \times 5)N \; KMnO_4 = 0.50N \; KMnO_4$$

In part (a) of the problem given, we require, as well, a similar conversion of the concentration of Fe^{2+}. Again, the oxidation state of Fe^{2+} changes from an initial $+2$ to a final of $+3$, for a total oxidation number increase of one unit; hence,

$$\frac{\text{mole } Fe^{2+}}{1} = 1 \text{ g-equiv of } Fe^{2+}, \text{ as a reductant}$$

and, consequently

$$\text{mole of } Fe^{2+} = 1 \text{ g-equiv of } Fe^{2+}, \text{ as a reductant.}$$

Correspondingly, because as a reductant

$$1M \; Fe^{2+} = 1N \; Fe^{2+}$$

then

$$0.20M \; Fe^{2+} = (0.20 \times 1)N \; Fe^{2+} = 0.20N \; Fe^{2+}.$$

Having thus acquired concentrations in the needed units, we may now proceed directly with the given problem:

(a) The direct general expression for use here is:

$$\frac{KMnO_4}{ml \times N} = \frac{Fe^{2+}}{ml \times N}.$$

Substitution of numerical values yields the required volume:

$$ml \times 0.50 = 30.0 \times 0.20$$

(b) Our set-up here is:

$$\underbrace{\frac{KMnO_4}{ml \times N}} = \underbrace{\frac{Fe^{2+}}{mequiv}}.$$

The given 30.0 mg of Fe^{2+} must be converted to milliequivalents. The atomic weight of Fe^0 being 55.8, there are then 55.8 mg of Fe^{2+} to the millimole; and, as for this reaction, there is one milliequivalent of Fe^{2+} ion to the millimole (in conformity with the loss of one electron), then, as a reductant,

$$\frac{55.8 \text{ mg/mmole}}{1 \text{ mequiv/mmole}} = 55.8 \text{ mg of } Fe^{2+}/\text{mequiv.}$$

Consequently, with 55.8 mg of Fe^{2+} constituting a single milliequivalent thereof, 30.0 mg would provide, for reducing purposes,

$$\frac{30.0 \text{ mg}}{55.8 \text{ mg/mequiv}} = 0.54 \text{ mequiv of } Fe^{2+}.$$

Conversion of a given weight of a species, be it in milligrams or grams, to required milliequivalents or gram-equivalents, respectively, is then nothing more than the simple division of the given weight by the corresponding molar weight which itself is divided by the total change in oxidation number involved. Appropriate substitution, numerically, for the terms of the original general expression yields the required calculable volume,

$$ml \times 0.50 = \frac{30.0}{55.8/1}.$$

(c) In conformity with its loss of one electron by oxidation,

$$\frac{mmole \ Fe^{2+}}{1} = 1 \text{ mequiv of } Fe^{2+}, \text{ as a reductant.}$$

Hence, as a reductant,

$$30.0 \text{ mmoles of } Fe^{2+} = 30.0 \text{ mequiv of } Fe^{2+}.$$

We may then substitute, as appropriate, to obtain the volume sought,

$$\underbrace{\frac{KMnO_4}{ml \times N}} = \underbrace{\frac{Fe^{2+}}{mequiv}}.$$

$$ml \times 0.50 = 30.0$$

The questions that perhaps most frequently present themselves in the experimental tasks incidental to theoretical evaluations of reaction stoichiometry involve the mathematics of preparing solutions, or of changing their concentrations to conform to prescribed procedures of work. These are, invariably, matters of solubilization and dilution, and largely without significant chemical change, excluding ionization and hydration for our present purposes. Without the chemical alterations that result from the transfers of electrons in oxidation-reduction, or from the realignments without impairing the charges of species involved in metathetical reaction, automatic conversions of other units of concentration to normality or to equivalents are hardly warranted. They are nonetheless correct, however, despite the time and energy that is expended unnecessarily. As total weight or other quantities of species that do not undergo chemical change must remain constant upon solubilization of a solute and its dilution thereafter, it is of distinct advantage in facilitating calculations for dilutions to bear in mind the following:

(a) Volume $\times$ concentration = total weight of solute.

Not only will the now-familiar expressions,

$$\text{liters} \times N = \text{total gram-equivalents}$$

and

$$\text{ml} \times N = \text{total milliequivalents}$$

apply, but also because volumes and concentrations are each expressible in a number of different units, all these, likewise validly conform to the given generalized expression; illustratively,

$$\text{liters} \times M = \text{total moles}$$

or

$$\text{ml} \times M = \text{total millimoles}$$

and

$$\text{liters} \times \text{grams/liter} = \text{total grams}$$

or

$$\text{ml} \times \text{mg/ml} = \text{total milligrams.}$$

(b) The reduction of the concentration of a solution by addition of volume-increasing nonreactant solvent cannot change the relationships of the specific solute common to both the more concentrated and the less concentrated solutions; that is,

$$\underbrace{\text{volume} \times \text{concentration}}_{\textit{before dilution}} = \underbrace{\text{volume} \times \text{concentration}}_{\textit{after dilution}}.$$

Therefore, our equation can be set up not only as heretofore:

$$\text{liters} \times N = \text{liters} \times N$$

or

$$\text{ml} \times N = \text{ml} \times N$$

but also,

$$\text{liters} \times M = \text{liters} \times M$$

or

$$\text{ml} \times M = \text{ml} \times M.$$

The designations of molarity in the preceding relationships may be directly used only in those cases wherein the solute does not undergo reaction and is being considered only with respect to preparation of its solution. Coincidental correctness will prevail only when the numerical values of M and N happen to be identical.

Likewise, for problems of dilution involving the reduction of the concentration of a dissolved solute by introducing additional solvent, we may write

$$\text{liters} \times \text{grams/liter} = \text{liters} \times \text{grams/liter}$$

or

$$\text{ml} \times \text{mg/ml} = \text{ml} \times \text{mg/ml}$$

and even

$$\text{liters} \times \% = \text{liters} \times \%$$

or

$$\text{ml} \times \% = \text{ml} \times \%.$$

As the stoichiometric calculations of solutions — their preparation, dilution, and reactions — make use of so many varied approaches, it is expedient to leave further denouements to the illustrative calculations that follow.

Sample Calculations

■ PROBLEM 1: To determine the concentration of a pure liquid from its density.

Calculate the formality (molarity) of H_2O in pure water at a temperature at which the density of the liquid is 0.9962 g/ml.

Solution.

$$1 \text{ ml of } H_2O \text{ weighs } 0.9962 \text{ g}$$
$$1 \text{ liter of } H_2O \text{ weighs } 996.2 \text{ g}$$

Inasmuch as one gram-formula weight of H_2O (= 18.0 g) represents one mole of H_2O, the concentration thereof is

$$\frac{996.2 \text{ g/liter}}{18.0 \text{ g/g-formula}} = 55.3 \text{ g-formulas/liter } H_2O$$

$$= 55.3F \ H_2O \ (= 55.3M).$$

■ PROBLEM 2: To derive the molar concentration of a species from the density of its solution.

Calculate the concentration of H_2O in 8.5 F H_2SO_4 at a temperature at which the density of the aqueous solution is 1.460 g/ml.

Solution. One gram-formula weight of solute, H_2SO_4, amounts to 98.1 g. Hence, one liter of an 8.5F solution of this solute provides

$$8.5 \text{ g-formula} \times 98.1 \text{ g/g-formula} = 834 \text{ g of solute } H_2SO_4.$$

The density of the solution ($= 1.460$ g/ml $= 1460$ g/liter) represents, however, the total weight of both solute and solvent. Consequently, in one liter of solution, the weight of H_2O is

$$1460 \text{ g} - 834 \text{ g} = 626 \text{ g}.$$

As one mole of H_2O is its gram-formula weight of 18.0 g, the concentration of H_2O in the solution is

$$\frac{626 \text{ g/liter}}{18.0 \text{ g/g-formula}} = 34.8 \text{ g-formulas/liter } H_2O$$

$$= 34.8F \; H_2O \; (= 34.8M).$$

■ PROBLEM 3: To derive the concentration formality of a species from the specific gravity of the solution.

Calculate the formality of an aqueous solution of H_2SO_4 having a specific gravity of 1.219 at 20°/4°C, and containing 30% by weight of anhydrous solute.

Solution. The specific gravity of the solution at the measured temperature of 20°C reveals that its density is 1.22 times greater than that of pure water at 4°C. As the density of H_2O at 4°C is accepted as 1.0 g/ml, the density of the solution is identical with its specific gravity; that is, 1.219 g/ml or 1219 g/liter. Of the total weight (solute + solvent), 30% is the weight of pure anhydrous solute alone; therefore,

$$1219 \text{ g/liter} \times 0.30 = 365.7 \text{ g/liter of pure solute } H_2SO_4.$$

As the g-formula weight of H_2SO_4 is 98.1 g, the formality of the solution

$$\frac{365.7 \text{ g/liter}}{98.1 \text{ g/g-formula weight}} = 3.7 \text{ g-formula weights/liter}$$

$$= 3.7F \; H_2SO_4.$$

■ PROBLEM 4: Applications of the different scales of concentration for expressing the quantity of dissolved solute in solution.

The density, at $t°C$, of a given aqueous solution of KCl containing 9.0 g of the salt in 100.0 g of solution is determined to be 1.063 g/ml. With respect to KCl as an ion-aggregate, express the concentration of this solution in terms of

(a) its molality.
(b) its formality.
(c) its mole-fraction of solute.

Solution.

(a) As defined with respect to the solute, the molality of a solution, m, constitutes the number of gram-formulas, or moles present in 1000m of solvent H_2O. As in a total of 100.0 g of solution the weight of KCl is only 9.0 g, the weight of H_2O making up the remainder of the required weight of solution will be:

$$100.0 \text{ g} - 9.0 \text{ g} = 91.0 \text{ g of } H_2O.$$

Hence, if 91.0 g of H_2O contain 9.0 g of dissolved KCl,

$$91.0 \text{ g of } H_2O \text{ contains } \frac{9.0 \text{ g KCl}}{74.56 \text{ g KCl/mole}} = 0.12 \text{ mole of KCl.}$$

Consequently,

$$1 \text{ g of } H_2O \text{ provides } 0.19/91.0 \text{ mole of KCl}$$

and

$$1000 \text{ g of } H_2O \text{ will provide } 1000 \times 0.12/91.0 = 1.32 \text{ mole of KCl.}$$

Our definition of molality is thus fulfilled; and the given weight of KCl (as formula units) represents a 1.32 m (m = molal) solution.

(b) The given solution is 9.0% by weight of solute KCl (9.0 g of solute in 100 g of solution). As the stated density of the solution requires that each milliliter thereof weigh a total (solvent + solute) of 1.063 g, clearly, the weight of KCl per liter of solution is

$$1.063 \frac{\text{g solution}}{\text{ml}} \times \frac{9 \text{ g KCl}}{100 \text{ g solution}} \times 1000 \frac{\text{ml}}{\text{liter}} = 95.62 \text{ g KCl/liter.}$$

The formality of the solution is therefore,

$$\frac{95.62 \text{ g KCl/liter}}{74.56 \text{ g KCl/g-formula}} = 1.28 \text{ g-formulas KCl/liter} = 1.28F \text{ KCl.}$$

(c) The mole-fraction of solute KCl (as an ion-pair or aggregate) in the given aqueous solution is obtained by substitution in the expression

$$\text{mole-fraction of solute} = \frac{\text{moles of solute}}{\text{moles of solute} + \text{moles of solvent}}.$$

The given 9.0%-by-weight solution supplies 9.0 g of solute KCl and 91.0 g of solvent H_2O for each 100.0 g of solution, conforming to

$$\frac{9.0 \text{ g KCl}}{74.56 \text{ g KCl/mole}} = 0.12 \text{ mole KCl}$$

and

$$\frac{91.0 \text{ g } H_2O}{18.02 \text{ g } H_2O/\text{mole}} = 5.05 \text{ mole } H_2O$$

whence, upon substitution, we have

$$\text{mole-fraction of KCl} = \frac{0.12 \text{ mole of KCl}}{0.12 \text{ mole of KCl} + 5.05 \text{ mole of } H_2O} = 0.023.$$

■ PROBLEM 5: To prepare a solution of specified concentration.

Calculate the weight, in grams, of solute required to make up 200 ml of an aqueous $0.80F$ solution of Na_2SO_4.

Solution. A $0.80F$ solution of Na_2SO_4 contains 0.80 g-formula weight of the dissolved salt in 1000 ml of solution. If

$$1 \text{ g-formula weight of } Na_2SO_4 = 142.05 \text{ g}$$

then

$$0.80 \text{ g-formula weight of } Na_2SO_4 = 0.80 \times 142.05 \text{ g} = 113.64 \text{ g}.$$

The solubilizing of this weight of solute within 1000 ml of total solution standardizes the make-up of a $0.80\,F$ solution of Na_2SO_4. We do not require, however, a full 1000 ml, but only 200 ml. If the needed volume is thus reduced to 200/1000 (one-fifth) of the calculated standard, the weight of the solute, Na_2SO_4, dissolved in this diminished volume must likewise be reduced to the same extent, if the concentration is to remain unchanged at $0.80\,F$. Hence, the required weight of solute, Na_2SO_4, is

$$113.64 \text{ g} \times 200 \text{ ml}/1000 \text{ ml} = 22.73 \text{ g}.$$

It is to be noted that this calculated weight of solute is not to be solubilized by taking 200 ml of solvent H_2O. Rarely will such a procedure yield with any particular solute the 200 ml of solution that might be sought, and which is definitely required here. Very significant expansion in volume frequently results during solubilization; and, at times, even contraction when hydration of the solute becomes extensive. The calculated weight of solute is first dissolved in some adequate volume of solvent H_2O that is less than the final volume of the solution being sought. When the solubilizing process is complete, the resultant solution is diluted with additional solvent, in a suitably-calibrated vessel (graduated cylinder, volumetric flask), to the ultimate volume required. This procedure is familiarly known as "filling to mark."

■ PROBLEM 6: To evaluate concentrations of species present in a solution containing a specific weight of dissolved solid.

Calculate the molar concentrations of solute species in a solution containing a dissolved weight of 21.56 g of $Al_2(SO_4)_3$ to 250 ml of solution. Assume the salt to be completely ionic in the solution, and also ignore hydrolysis effects.

Solution. In conformity with the stated conditions, the solid, once dissolved, is completely dissociated into ions whose individual concentrations are presumed to be unimpaired by side reactions. Hence, in accordance with the equation,

$$Al_2(SO_4)_{3\,(s)} \rightarrow 2Al^{3+} + 3SO_4^{2-}$$

solubilization must provide that for each gram-formula weight of $Al_2(SO_4)_3$ dissolved, there will be obtained two moles of Al^{3+} ion and three moles of SO_4^{2-} ion. Inasmuch as one gram-formula weight of the $Al_2(SO_4)_3$ amounts to 342.13 g, the given solution provides

$$\frac{25.66 \text{ g } Al_2(SO_4)_3}{342.13 \text{ g } Al_2(SO_4)_3/ \text{ g-formula}} = 0.075 \text{ g-formula}$$

of this solute in 250 ml. Therefore, there is present in the given volume of solution a total of

$$\frac{2 \text{ moles } Al^{3+}}{1 \text{ g-formula } Al_2(SO_4)_3} \times 0.075 \text{ g-formula } Al_2(SO_4)_3 = 0.15 \text{ mole } Al^{3+}$$

and

$$\frac{3 \text{ moles } SO_4^{2-}}{1 \text{ g-formula } Al_2(SO_4)_3} \times 0.075 \text{ g-formula } Al_2(SO_4)_3 = 0.225 \text{ mole } SO_4^{2-}.$$

We are asked, however, for the molar concentration, which, by definition, is established as moles/liter; consequently,

$$\frac{\text{moles}}{\text{liter}} Al^{3+} = 0.15 \text{ mole } Al^{3+} \times \frac{1000 \text{ ml/liter}}{250 \text{ ml}}$$

$$= 0.60 \text{ mole/liter } Al^{3+} = 0.60M \text{ } Al^{3+}$$

and

$$\frac{\text{moles}}{\text{liter}} SO_4^{2-} = 0.225 \text{ mole } SO_4^{2-} \times \frac{1000 \text{ ml/liter}}{250 \text{ ml}}$$

$$= 0.90 \text{ mole/liter } SO_4^{2-} = 0.90M \text{ } SO_4^{2-}.$$

Complete dissociation having been assumed, there are zero moles/liter of $Al_2(SO_4)_3$ as such in the solution. Clearly, then, the formality of the solution, (which implies nothing with respect to any alterations through solubilization in the chemical identity of the parent solute, assumed or otherwise) will be

$$0.075 \text{ g-formula } Al_2(SO_4)_3 \times \frac{1000 \text{ ml/liter}}{250 \text{ ml}}$$

$$= 0.30 \text{ g-formula/liter } Al_2(SO_4)_3 = 0.30F \; Al_2(SO_4)_3,$$

conforming, as observed, to the stoichiometry of the equation of the reaction.

■ PROBLEM 7:   To prepare a solution containing a specific quantity of an ion from a salt of that ion.

Calculate the weight in grams of $BaCl_2$ required to be dissolved in water in order to make up 60.0 ml of solution of concentration 12 mg of Ba^{2+} ion to a milliliter.

Solution. We require a total of

$$60.0 \text{ ml} \times 12 \; \frac{\text{mg Ba}^{2+}}{\text{ml}} = 720 \text{ mg Ba}^{2+} = 0.72 \text{ g Ba}^{2+}.$$

One gram-formula weight of the completely ionic $BaCl_2$ ($= 208.25$ g) supplies, upon dissolution, one gram-ion or mole of Ba^{2+} ($= 137.34$ g). Consequently, as 137.34 g of Ba^{2+} ion derives from 208.25 g of $BaCl_2$,

$$1 \text{ g of Ba}^{2+} \text{ ion derives from } 208.25/137.34 \text{ g of BaCl}_2$$

and

$$0.72 \text{ g of Ba}^{2+} \text{ ion derives from}$$

$$0.72 \times 208.25/137.34 \text{ g of BaCl}_2 = 1.09 \text{ g of BaCl}_2.$$

This weight, when dissolved in sufficient water to make 60.0 ml of solution, provide the required quantity of 12 mg of Ba^{2+} ion per ml.

The *factor-unit* method is always highly useful, either as a direct approach to the solution of a problem in stoichiometry or as a check upon the correctness of an answer otherwise derived. Thus, in the foregoing problem, the stepwise delineations could have been collated in the following manner. The required total grams of Ba^{2+} ion for 60.0 ml of solution is expressed by:

$$\frac{12 \text{ mg Ba}^{2+}}{\text{ml}} \times 60.0 \text{ ml} \times \frac{1 \text{ g}}{1000 \text{ mg}} = 0.72 \text{ g Ba}^{2+}$$

and the required total grams of $BaCl_2$ for 60.0 ml of solution is expressed by

$$\frac{208.25 \text{ g BaCl}_2}{1 \text{ g-formula BaCl}_2} \times \frac{1 \text{ g-formula BaCl}_2}{1 \text{ g-ion Ba}^{2+}} \times \frac{1 \text{g-ion Ba}^{2+}}{137.34 \text{ g Ba}^{2+}} \times 0.72 \text{ g Ba}^{2+}$$

$$= 1.09 \text{ g BaCl}_2.$$

As pursued, the dimensional arrangements of all terms must be such that unwanted species and units cancel out from the ultimate mathematical expressions and leave only the species and units that are being sought.

———————

■ PROBLEM 8: To reduce the concentration of a prepared solution by diluting its volume.

(a) Calculate the volume to which 20.0 ml of a $0.90F$ solution of $AgNO_3$ must be diluted in order to make up a $0.60F$ solution of the salt.

(b) Calculate the volume, in milliliters, of a $2.5F$ solution of $Cu(NO_3)_2$ that must be taken to yield, upon dilution with water, 85.0 ml of a solution containing 15 mg of Cu^{2+} ion to the milliliter.

Solution. For both parts we use the "volume-concentration" relationship

$$\underbrace{\text{volume} \times \text{concentration}}_{\substack{\text{more concentrated} \\ \text{solution}}} = \underbrace{\text{volume} \times \text{concentration}}_{\substack{\text{more dilute} \\ \text{solution}}}.$$

(a) If x = total ml of solution after the required dilution, appropriate substitution yields

$$20.0 \text{ ml} \times 0.90F = x \times 0.60F$$

whence

$$x = 18.0/0.60 = 30 \text{ ml (allowance made for significant figures).}$$

Mere inspection, itself, would have led quickly to the same result as obtained by the formalistic approach just used. As the concentration is being diminished to $0.60/0.90 (=\frac{2}{3})$ of its original value, the original volume must clearly increase in exactly the inverse proportion; that is, to $0.90/0.60$ or $\frac{3}{2}$ times the original. Hence,

$$\tfrac{3}{2} \times 20.0 \text{ ml} = 30 \text{ ml, as before.}$$

(b) As all units must be expressed alike for correct substitution in the volume-concentration relationship, we must either convert $2.5F$ $Cu(NO_3)_2$ to its representative number of milligrams of Cu^{2+} ion/ml, or convert 15 mg of Cu^{2+} ion/ml to its equivalent in terms of the formality of $Cu(NO_3)_2$ as salt. (To elect the former: the given $2.5F$ $Cu(NO_3)_2$ solution is $2.5M$ in Cu^{2+} ion, $5.0M$ in NO_3^- ion). As a $1M$ Cu^{2+} ion solution would contain 63.54 g of Cu^{2+}/liter *or* 63.54 mg/ml, the solution that has been provided contains

$$2.5 \times 63.54 \text{ mg } Cu^{2+}/\text{ml} = 158.85 \text{ mg } Cu^{2+}/\text{ml.}$$

Therefore, upon appropriate substitution in the volume-concentration relationship wherein

x = number of milliliters of $2.5F$ $Cu(NO_3)_2$ solution to be

taken (= solution of concentration 158.85 mg of Cu^{2+}/ml)

we have

$x \times 158.85$ mg of Cu^{2+}/ml = 85.0 ml $\times$ 15 mg of Cu^{2+}/ml

whence

$x = 8.0$ ml of $Cu(NO_3)_2$ solution.

Were we to elect, instead, to convert the solution of 15 mg of Cu^{2+}/ml to its corresponding molarity, then (inasmuch as a $1M$ Cu^{2+} ion solution must contain 63.54 mg of Cu^{2+}/ml) it follows that a solution of this cation containing less than this weight of it to the milliliter must be less than $1M$, and one that contains more than this weight to the milliliter must be more than $1M$. Our diluted solution here is, therefore, $15/63.54$ molar. With the value of x representing the required number of milliliters of the $2.5M$ solution of Cu^{2+} ion (2.5 F $Cu(NO_3)_2$), appropriate substitution within the volume-concentration relationship yields

$x \times 2.5M$ Cu^{2+} = 85.0 ml $\times$ $(15/63.54)M$ Cu^{2+}

whence

$x = 8.0$ ml of $Cu(NO_3)_2$ solution, as before.

■ PROBLEM 9: To determine the volumes of different solutions of the same solute, required to achieve by admixture a specific common concentration.

Calculate the respective volumes, in milliliters, of $0.35F$ KNO_3 and of $1.25F$ KNO_3 that, when mixed without extraneous addition of H_2O, will supply 85.0 ml of a $0.60F$ solution of KNO_3. Assume the admixed volumes to be exactly additive.

Solution. To be kept in mind is the significance of the volume-concentration expression. When it is applied in the form,

$$ml \times F = ml \times F$$

we are stressing the numerical equality of the total number of formula weights of KNO_3 that must exist upon both sides of the relationship. The total number of formula weights conforms to the equalities:

total gram-formula weights = total liters $\times$ formality of solution

or

total milligram-formula weights = total milliliters $\times$ formality of solution.

Clearly, then, it is inescapable that upon mixing

$$\left[\begin{array}{l}\text{gram-formulas from } 0.35F \text{ KNO}_3 \\ + \text{ gram-formulas from } 1.25F \text{ KNO}_3\end{array}\right] = \begin{array}{l}\text{gram-formulas of KNO}_3 \text{ of} \\ mixture\end{array}$$

or

$$\left[\begin{array}{l}\text{milligram-formulas from } 0.35F \text{ KNO}_3 \\ + \text{ milligram-formulas from } 1.25F \text{ KNO}_3\end{array}\right] = \begin{array}{l}\text{milligram-formulas of} \\ \text{KNO}_3 \text{ of } mixture.\end{array}$$

Hence, our substitution in the equated relationship,

$$\underbrace{\text{ml} \times F}_{\substack{\text{mg-formulas from} \\ 0.35F \text{ KNO}_3}} + \underbrace{\text{ml} \times F}_{\substack{\text{mg-formulas from} \\ 1.25F \text{ KNO}_3}} = \underbrace{\text{ml} \times F}_{\substack{\text{mg-formulas from} \\ mixture}}$$

yield to the following considerations:

Let x = milliliters of the $0.35F$ solution taken;

whereupon, because, as required, 85.0 ml must be the total volume of the ultimate mixture being prepared,

$85.0 - x$ = milliliters of the $1.25F$ solution taken

consequently, the volume-concentration relationship conforms to the numerical values of the two different solutions, and of the mixture sought, as follows:

$$(x \times 0.35F) + (85.0 - x) \times 1.25F = 85.0 \text{ ml} \times 0.60F$$

Mathematical resolution of this expression thereupon yields

$$x = 61.4 \text{ ml of } 0.35F \text{ KNO}_3 \text{ solution required}$$

and

$$(85.0 - x) = 23.6 \text{ ml of } 1.25F \text{ KNO}_3 \text{ solution required.}$$

■ PROBLEM 10: To determine the quantities and concentrations of species remaining after metathetical (nonredox) reaction.

To 20.0 ml of $0.30F$ H$_2$SO$_4$ are added 5.0 ml of $2.0F$ KOH. For each of the different solute species actually present after the ensuing neutralization calculate its molarity; its metathetical normality; its total number of millimoles; its total number of milliequivalents; its millimoles to milliliter; its milliequivalents to milliliter.

Solution. Reaction proceeds in accordance with the neutralization,

$$\text{H}^+ + \text{OH}^- \rightarrow \text{H}_2\text{O}.$$

The $0.30F$ solution of H$_2$SO$_4$ before reaction provides

$$0.60M \text{ H}^+ \text{ ion and } 0.30M \text{ SO}_4^{2-} \text{ ion}$$

assuming 100% ionization of the solute by

$$H_2SO_4 \rightarrow 2H^+ + SO_4^{2-}.$$

The 2.0F solution of KOH before reaction provides

2.0M K$^+$ ion and 2.0M OH$^-$ ion.

We have no molecular KOH; and we ignore ion-aggregation. Utilizing the volume-concentration relationship in the form

$$\frac{\text{milliequivalents}}{\text{milliliters} \times \text{normality}} \quad \frac{\text{milliequivalents}}{\text{milliliters} \times \text{normality}},$$

we determine how many total milliequivalents we have of each of the species initially present in the separate and distinct solutions before mixing:

1. with respect to H$^+$ ion,

2. 0.30M H$^+$ = 0.30N H$^+$ = 0.30 mequiv H$^+$/ml.

Hence we have initially a total of

20.0 ml $\times$ 0.30 mequiv/ml H$^+$ = 6.0 mequiv of H$^+$.

2. With respect to OH$^-$ ion,

2.0M OH$^-$ = 2.0N OH$^-$ = 2.0 mequiv OH$^-$/ml.

Hence we have initially a total of

5.0 ml $\times$ 2.0 mequiv/ml OH$^-$ = 10.0 mequiv of OH$^-$.

3. With respect to K$^+$ ion,

2.0M K$^+$ = 2.0N K$^+$ = 2.0 mequiv K$^+$/ml.

Hence we have, initially, thereof a total of

5.0 ml $\times$ 2.0 mequiv/ml K$^+$ = 10.0 mequiv K$^+$.

4. With respect to SO$_4^{2-}$ ion,

0.30M SO$_4^{2-}$ = 0.60N SO$_4^{2-}$ = 0.60 mequiv SO$_4^{2-}$/ml.

Hence we have initially a total of

20.0 ml $\times$ 0.60 mequiv/ml SO$_4^{2-}$ = 12.0 mequiv of SO$_4^{2-}$.

The metathetical reaction from mixing the solutions alters the *total* quantities of only the H$^+$ and the OH$^-$ ions of the four species with which we are involved. The K$^+$ and SO$_4^{2-}$ ions are merely "spectators" to the chemical reaction, although their concentrations will have changed upon mixing of the two solutions. As we have introduced 6.0 mequiv of H$^+$ ion and 10.0 mequiv of OH$^-$ ion, the entire smaller quantity of H$^+$ ion will have been destroyed by neutralization together with an identical quantity of OH^- ion. We then have, after reaction, an excess of OH^- ion remaining free and unneutralized; that is

$$10.0 \text{ mequiv } OH^- - 6.0 \text{ mequiv } OH^- = 4.0 \text{ mequiv } OH^-.$$
$$\text{(initially)} \qquad \text{(used up)} \qquad \text{(excess)}$$

We are, however, also required to determine the concentrations (molarity and normality) of the species. Consequently, we must consider the total volume of the mixture which, assuming that the reactant volumes are exactly additive, is 25.0 ml (20.0 ml of H_2SO_4 plus 5.0 ml of KOH). Hence, the final solution is

1. with respect to concentration of H^+ ion,
   negligibly small and, in any event, dependent upon the molarity of the excess of OH^- ion presently to be evaluated.
2. with respect to concentration of OH^- ion,

$$4.0 \text{ mequiv } OH^-/25.0 \text{ ml} = 0.16N \; OH^- = 0.16M \; OH^-.$$

The actual concentration of the H^+ ion in the final mixture is now ascertainable by dividing this value of OH^- ion into the *ion-product* constant of water (1.0×10^{-14} at room temperature). This constant defines mathematically the mutual tolerances or compatibilities of H^+ and OH^- ions in any common aqueous solution and is obtained by multiplying the concentrations, in moles/liter, of these ions.

3. With respect to concentration of K^+ ion,

$$10.0 \text{ mequiv } K^+/25.0 \text{ ml} = 0.40N \; K^+ = 0.40M \; K^+$$

4. With respect to concentration of SO_4^{2-} ion,

$$12.0 \text{ mequiv } SO_4^{2-}/25 \text{ ml} = 0.48N \; SO_4^{2-} = 0.24M \; SO_4^{2-}$$

Collecting our several results we have

$$\begin{array}{r} M \text{ and } N \\ \text{after mixing} \end{array} \left\{ \begin{array}{l} SO_4^{2-} = 0.24M = 0.48N \\ K^+ = 0.40M = 0.40N \\ OH^- = 0.16M = 0.16N \end{array} \right.$$

$$\begin{array}{r} \text{mmole/ml and mequiv/ml} \\ \text{after mixing} \end{array} \left\{ \begin{array}{l} SO_4^{2-} = 0.24 \text{ mmole/ml} = 0.48 \text{ mequiv/ml} \\ K^+ = 0.40 \text{ mmole/ml} = 0.40 \text{ mequiv/ml} \\ OH^- = 0.16 \text{ mmole/ml} = 0.16 \text{ mequiv/ml} \end{array} \right.$$

$$\begin{array}{r} \text{Total mmole} \\ \text{after mixing} \\ [(\text{mmole/ml}) \times 25.0 \text{ ml}] \\ \text{Total mequiv} \\ \text{after mixing} \\ [(\text{mequiv/ml}) \times 25.0 \text{ ml}] \end{array} \left\{ \begin{array}{l} SO_4^{2-} = 0.60 \text{ mmole} = 12.0 \text{ mequiv} \\ K^+ = 10.0 \text{ mmole} = 10.0 \text{ mequiv} \\ OH^- = 16.0 \text{ mmole} = 16.0 \text{ mequiv.} \end{array} \right.$$

■ PROBLEM 11: To determine the molar concentrations of solute species by utilizing specific gravity and weight-percent of solution.

The label on a bottle of HNO_3 solution reads "69.8% by weight; specific gravity 1.42." Calculate the concentration formality of this acid.

Solution. We obtain the formality of the solution by comparing its concentration with that of the standard that represents a $1F$ solution of HNO_3. Inasmuch as a solution that is $1F$ in HNO_3 must contain 63.0 g (one gram-formula weight) of pure HNO_3 solute to the liter, if the given solution contains more than 63.0 g of pure HNO_3 solute it will be more than $1 F$; and if it contains less than this weight of the solute it will be less than $1 F$.

What, then, does the given solution contain in weight of this solute? As the specific gravity of the solution is 1.42, we accept that each milliliter of it weighs 1.42 g. This is the weight of both solute and solvent that make up the milliliter of the solution. The percent by weight that constitutes the contribution of the solute HNO_3 to the total weight of the solution is 69.8%, as stated. Hence, the simple multiplication that expresses the weight of solute HNO_3 in a liter of solution is

$$1.42 \; \frac{\text{g solution}}{1 \text{ ml solution}} \times \frac{1000 \text{ ml solution}}{1 \text{ liter solution}} \times \frac{0.698 \text{ g pure } HNO_3 \; solute}{1 \text{ g solution}}$$

$$= 991.16 \text{ g pure } HNO_3 \; solute/1 \text{ liter solution.}$$

The concentration of the given solution is greater than $1F$; that is

$$\frac{991.6 \text{ g } HNO_3}{\text{liter solution}} \times \frac{1 \text{ g-formula } HNO_3}{63.0 \text{ g } HNO_3}$$

$$= 15.7 \text{ g-formulas } HNO_3/\text{liter solution} = 15.7F \; HNO_3.$$

■ PROBLEM 12: To determine the equivalent weight of a solute to be used under specified reaction conditions.

Calculate the gram-equivalent weight of HNO_3 as used in each of the following reactions:

(a) Solely for its acidic content.
(b) As an oxidant, undergoing reduction to NO^0 gas.
(c) As an oxidant, undergoing reduction to NO_2 gas.
(d) As an oxidant, undergoing reduction to NH_4^+ ion.

Solution.

(a) As an *acid*, the equivalent weight of HNO_3 is that weight that supplies in reaction, on demand by a base, one mole of H^+ ion (one gram-equivalent). Therefore, as one mole of HNO_3 can yield one mole of H^+ ion, the gram-equivalent weight of HNO_3 functioning as an acid is

$$63.0 \; \frac{\text{g}}{1 \text{ g-formula}} \times \frac{1 \text{ g-formula}}{1 \text{ g-equiv}} = 63 \; \frac{\text{g}}{1 \text{ g-equiv}}$$

(b) As an *oxidant*, the gram-equivalent weight of HNO_3 is that weight that accepts one Avogadro number of electrons from a reductant; that is, it accepts the gram-electron. As it is the nitrate ion, NO_3^-, that is the specific oxidizing constituent of the HNO_3^-, we determine the required gain in electrons either directly from the change in oxidation number of the element nitrogen in the NO_3^- ion as it goes from its initial to its final state; or from the balanced ion-electron half-reaction (the "partial"). In the NO_3^- ion the oxidation state of the nitrogen atom is $+5$; in the reaction product, NO^0, the oxidation state of the nitrogen atom is $+2$. Consequently, the latter represents an over-all decrease in the oxidation number of nitrogen of $+3$ in a mole of NO_3^- ion, which is equivalent to a gain of three electrons.

The balanced partial for the reduction, had we elected such an approach, must lead to the same result:

$$NO_3^- + 4H^+ + 3e^- \rightarrow NO^0 + 2H_2O.$$

The equivalent weight of the HNO_3 as an oxidant undergoing reduction to NO^0 is therefore

$$63.0 \; \frac{g}{1 \text{ g-formula}} \times \frac{1 \text{ g-formula}}{3 \text{ g-equiv}} = 21.0 \text{ g/g-equiv.}$$

(c) In the reduction of the NO_3^- ion to NO_2 gas, the oxidation state of the nitrogen changes from an initial $+5$ to a final $+4$. The over-all decrease in oxidation number with respect to the element nitrogen in a mole of nitrate ion is $+1$; or, (had we elected such an approach) equivalent to a gain of one electron as similarly obtained from inspection of the balanced partial:

$$NO_3^- + 2H^+ + 1e^- \rightarrow NO_2 + H_2O.$$

Hence, the equivalent weight of HNO_3 as an oxidant undergoing reduction to NO_2 is therefore:

$$63.0 \; \frac{g}{1 \text{ g-formula}} \times \frac{1 \text{ g-formula}}{1 \text{ g-equiv}} = 63.0 \text{ g/g-equiv.}$$

(d) In the reduction of the NO_3^- ion to NH_4^+ ion, the oxidation state of nitrogen changes from an initial $+5$ to a final -3. The over-all decrease in oxidation number in nitrogen to a mole of the nitrate ion is, therefore, $+8$, which is equivalent to the gain of eight electrons as similarly revealed by the pertinent balanced partial (when preferred as an approach):

$$NO_3^- + 10H^+ + 8e^- \rightarrow NH_4^+ + 3H_2O.$$

Consequently, the equivalent weight of HNO_3 as an oxidant undergoing reduction to NH_4^+ ion is

$$63.0 \; \frac{g}{1 \text{ g-formula}} \times \frac{1 \text{ g-formula}}{8 \text{ g-equiv}} = 7.9 \text{ g/g-equiv.}$$

■ PROBLEM 13: Application of the volume-concentration expression to the variable ionizations of a polyprotic acid in neutralization.

Calculate the weight in grams of NaOH required for each of the following reactions:

(a) 0.016 liter of $0.50F$ H_3PO_4 to (1) $H_2PO_4^-$ ion; (2) HPO_4^{2-} ion; (3) PO_4^{3-} ion

(b) 0.016 g of H_3PO_4 to (1) $H_2PO_4^-$ ion; (2) HPO_4^{2-} ion; (3) PO_4^{3-} ion

(c) 0.016 g-equivalents of H_3PO_4 to (1) $H_2PO_4^-$ ion; (2) HPO_4^{2-} ion; (3) PO_4^{3-} ion.

Solution.

(a) We bear in mind the variabilities possible in the gram-equivalent weights of a polyprotic acid. The given acid, orthophosphoric acid (H_3PO_4), has three different equivalent weights, each a function of the differing demands made upon it for its content of H^+ ion by the added base, NaOH. Thus,

$$H_3PO_4 + OH^- \rightarrow H_2O + H_2PO_4^-$$

$$H_3PO_4 + 2OH^- \rightarrow 2H_2O + HPO_4^{2-}$$

$$H_3PO_4 + 3OH^- \rightarrow 3H_2O + PO_4^{3-}$$

for which, in the sequence given, the gram-equivalent weights of the H_3PO_4 are, respectively, one gram-formula weight, one-half the gram-formula weight, and one-third the gram-formula weight.

We proceed to the appropriate substitutions in the applicable volume-concentration relationship for reaction; that is

$$\underbrace{\frac{\text{gram-equivalents of NaOH}}{\text{liters} \times \text{normality}}} \quad \underbrace{\frac{\text{gram-equivalents of } H_3PO_4}{\text{liters} \times \text{normality}}}.$$

We equate $0.50F$ $H_3PO_4 \approx 0.50M$ H_3PO_4 in the over-all aqueous triprotic equilibria of its solution; and, for its required conversions:

(1) to $H_2PO_4^-$ ion: $0.50M$ $H_3PO_4 = 0.50N$ H_3PO_4
$$= 0.50 \text{ g-equiv/liter,}$$

(2) to HPO_4^{2-} ion: $0.50M$ $H_3PO_4 = 1.0N$ H_3PO_4
$$= 1.0 \text{ g-equiv/liter,}$$

(3) to PO_4^{3-} ion: $0.50M$ $H_3PO_4 = 1.5N$ H_3PO_4
$$= 1.5 \text{ g-equiv/liter.}$$

Hence, where $x =$ required weight of NaOH,

(a.1) $$\frac{x}{\dfrac{40.0 \text{ g NaOH}}{1 \text{ g-formula}} \times \dfrac{1 \text{ g-formula NaOH}}{1 \text{ g-equiv}}} = 0.016 \text{ liter } H_3PO_4$$

$$= \frac{0.50 \text{ g-equiv}}{1 \text{ liter}} H_3PO_4$$

whence, $x = 0.32$ g NaOH, required to convert to $H_2PO_4^-$ ion.

(a.2) $\dfrac{x}{\dfrac{40.0 \text{ g NaOH}}{1 \text{ g-formula}} \times \dfrac{1 \text{ g-formula NaOH}}{1 \text{ g-equiv}}} = 0.016 \text{ liter } H_3PO_4$

$$\times \frac{1.0 \text{ g-equiv}}{1 \text{ liter}} H_3PO_4$$

whence, $x = 0.64$ g NaOH, required to convert to HPO_4^{2-}.

(a.3) $\dfrac{x}{\dfrac{40.0 \text{ g NaOH}}{1 \text{ g-formula}} \times \dfrac{1 \text{ g-formula NaOH}}{1 \text{ g-equiv}}} = 0.016 \text{ liter } H_3PO_4$

$$= \frac{1.5 \text{ g-equiv}}{1 \text{ liter}} H_3PO_4$$

whence $x = 0.096$ g NaOH, required to convert to PO_4^{3-} ion.

(b) Following the same pattern as in the preceding part, we set

$$x = \text{required weight of NaOH,}$$

whereupon

(b.1) $\dfrac{x}{\dfrac{40.0 \text{ g NaOH}}{1 \text{ g-formula}} \times \dfrac{1 \text{ g-formula NaOH}}{1 \text{ g-equiv}}}$

$$= \frac{0.016 \text{ g } H_3PO_4}{\dfrac{98.0 \text{ g } H_3PO_4}{1 \text{ g-formula}} \times \dfrac{1 \text{ g-formula } H_3PO_4}{1 \text{ g-equiv}}}$$

whence $x = 0.0065$ g NaOH, required to convert to $H_2PO_4^-$ ion.

(b.2) $\dfrac{x}{\dfrac{40.0 \text{ g NaOH}}{1 \text{ g-formula}} \times \dfrac{1 \text{ g-formula NaOH}}{1 \text{ g-equiv}}}$

$$= \frac{0.016 \text{ g } H_3PO_4}{\dfrac{98.0 \text{ g } H_3PO_4}{1 \text{ g-formula}} \times \dfrac{1 \text{ g-formula } H_3PO_4}{2 \text{ g-equiv}}}$$

whence $x = 0.013$ g NaOH, required to convert to HPO_4^{2-} ion.

(b.3) $\dfrac{x}{\dfrac{40.0 \text{ g NaOH}}{1 \text{ g-formula}} \times \dfrac{1 \text{ g-formula NaOH}}{1 \text{ g-equiv}}}$

$$= \frac{0.016 \text{ g } H_3PO_4}{\dfrac{98.0 \text{ g } H_3PO_4}{1 \text{ g-formula}} \times \dfrac{1 \text{ g-formula } H_3PO_4}{3 \text{ g-equiv}}}$$

whence $x = 0.020$ g NaOH, required to convert to PO_4^{3-} ion.

(c) The significance of the volume-concentration expression is inescap-ably that in chemical reaction identical numbers of equivalents of the different reactants must take part, regardless of the identities of the chemical products into which they are converted. In all instances, then, whether converted to $H_2PO_4^-$, HPO_4^{2-}, or PO_4^{3-}, the given 0.016 g-equiv of H_3PO_4 must interact with 0.016 g-equiv of the NaOH. The actual weights that constitute 0.016 g-equiv of H_3PO_4 in the respective conversions to each of the three *different* possible products, which interact with 0.016 g-equiv of NaOH, is different. This is inherent in the concept of the gram-equivalent. Hence, with x = required weight of NaOH we derive

$$\frac{x \text{ g NaOH}}{40.0 \text{ g/g-equiv}} = 0.016 \text{ g-equiv NaOH}$$

$$x = 0.64 \text{ g NaOH}$$

required to convert the identical given numbers of g-equiv of H_3PO_4 to $H_2PO_4^-$, or to HPO_4^{2-}, or to PO_4^{3-}.

■ PROBLEM 14:   To determine the volume of a reactant required for a specific use.

Calculate in each instance the volume, in milliliters, of reagent $0.30M$ $Cr_2O_7^{2-}$ ion required to

(a) precipitate 100 mg of Pb^{2+} ion by the reaction,

$$2Pb^{2+} + Cr_2O_7^{2-} + H_2O \rightarrow 2PbCrO_4 + 2H^+.$$

(b) oxidize 5.0 millimoles of Fe^{2+} ion in acidic medium by the reaction,

$$6Fe^{2+} + Cr_2O_7^{2-} + 14H^+ \rightarrow 6Fe^{3+} + 2Cr^{3+} + 7H_2O.$$

(c) match exactly the oxidizing capacity of 6.4 ml of $0.30M$ MnO_4^- ion. undergoing conversion in acidic medium in accordance with the equation

$$5Fe^{2+} + MnO_4^- + 8H^+ \rightarrow 5Fe^{3+} + Mn^{2+} + 4H_2O.$$

Solution. The volume of the reagent $Cr_2O_7^{2-}$ ion needed in each instance is determined by its number of equivalents involved in the respective reaction. It must always be equal to the number of equivalents of the species with which it reacts.

(a) Here we are concerned with a metathetical (non-redox) reaction, because neither the Pb^{2+} ion nor the Cr atom in the dichromate ion undergoes any alteration in its initial oxidation state. That of Pb^{2+} remains $+2$; that of the Cr atom in its respective polyatomic ion remains $+6$, both initially and in their common reaction product, $PbCrO_4$. Consequently, to determine the equivalent weight of either is not a matter of electron transfer, but rather of the total valence of a mole of the reacting ions. As one mole of $Cr_2O_7^{2-}$ ion in this instance contains two equivalents,

$$1M \ Cr_2O_7^{2-} = 2N Cr_2O_7^{2-} = 2 \ \textit{metathetical} \ \text{mequiv/ml of } Cr_2O_7^{2-} \text{ ion}$$

and

$$0.30M \ Cr_2O_7^{2-} = 0.60N \ Cr_2O_7^{2-}$$
$$= 0.60 \ metathetical \ mequiv/ml \ of \ Cr_2O_7^{2-} \ ion.$$

Letting x = milliliters of $Cr_2O_7^{2-}$ ion, our substitution in the expression

$$\frac{\overbrace{Cr_2O_7^{2-}}}{ml \times N} = \frac{\overbrace{Pb^{2+}}}{mequiv}$$

yields

$$x \times \frac{0.60 \ mequiv}{1 \ ml} \ Cr_2O_7^{2-} = \frac{100 \ mg \ Pb^{2+}}{\dfrac{207.2 \ mg \ Pb^{2+}}{1 \ mmole}} \times \frac{1 \ mmole \ Pb^{2+}}{2 \ mequiv}$$

whence

$x = 1.6$ ml $Cr_2O_7^{2-}$ ion solution, required to precipitate $PbCrO_4$.

(b) We are involved here with a redox reaction, for which the equivalent weight of $Cr_2O_7^{2-}$, as the oxidant, is determined by the total number of electrons it accepts for a mole in being reduced. The oxidation state of the Cr atom changes from $+6$ in the $Cr_2O_7^{2-}$ ion to $+3$ in its reduction product, Cr^{3+}. Hence, for the $Cr_2O_7^{2-}$ ion, the total decrease in oxidation number is $+6$; and this conforms, as would be expected, to the information offered by the balanced half-reaction (when preferred as an approach).

$$Cr_2O_7^{2-} + 14H^+ + 6e^- \rightarrow 2Cr^{3+} + 7H_2O.$$

The equivalent of the $Cr_2O_7^{2-}$ ion is consequently one-sixth of the mole of it; therefore,

$$1M \ Cr_2O_7^{2-} = 6N \ Cr_2O_7^{2-} = 6 \ mequiv/ml \ Cr_2O_7^{2-}$$

and

$$0.30M \ Cr_2O_7^{2-} = 1.8N \ Cr_2O_7^{2-} = 1.8 \ mequiv/ml \ Cr_2O_7^{2-}.$$

With respect now to the oxidation-conversion of the Fe^{2+} ion to the Fe^{3+} ion, an increase of $+1$ in oxidation state is noted similarly to be determined (when preferred) from the half-reaction

$$Fe^{2+} \rightarrow Fe^{3+} + 1e^-.$$

Therefore, as

$$1 \ mmole \ Fe^{2+} = 1 \ mequiv \ Fe^{2+}$$

and, with x = milliliters of the solution of $Cr_2O_7^{2-}$ ion required, we may substitute in the volume-concentration relationship

$$x \times \frac{1.8 \ mequiv}{ml} \ Cr_2O_7^{2-} = \frac{5.0 \ mmoles \ Fe^{2+}}{1 \ mmole/mequiv \ Fe^{2+}}$$

whence

$x = 2.8$ ml of the $Cr_2O_7^{2-}$ ion solution, required for oxidation of Fe^{2+}.

(c) In order for the solution of $Cr_2O_7^{2-}$ ion to match the oxidizing capacity of the MnO_4^- ion solution, it must provide the same number of equivalents as does the latter. Because in the reduction conversion of MnO_4^- ion to Mn^{2+} ion, the manganese undergoes a decrease of $+5$ in oxidation number (acceptance of five electrons by the mole), and as similarly derived from the partial,

$$MnO_4^- + 8H^+ + 5e^- \rightarrow Mn^{2+} + 4H_2O$$

the following relationship holds —

$$1M \; MnO_4^- = 5N \; MnO_4^- = \frac{5 \; mequiv}{ml} \; MnO_4^-$$

and, consequently,

$$0.30M \; MnO_4^- = 1.5N \; MnO_4^- = \frac{1.5 \; mequiv}{ml} \; MnO_4^-.$$

Inasmuch as we start with 6.4 ml of the $1.5N$ MnO_4^- solution, it follows that

Total mequiv $MnO_4^- = 6.4$ ml $\times$ 1.5 mequiv/ml MnO_4^-
$$= 9.6 \; mequiv \; MnO_4^-.$$

Clearly, it is required that the $Cr_2O_7^{2-}$ ion match this with an identical number of milliequivalents. We have already ascertained (in part b) that the given $0.30M$ solution of $Cr_2O_7^{2-}$ ion under the same conditions of redox is $1.8N$ ($=1.8$ mequiv/ml). Therefore, if

$$x = ml \; of \; Cr_2O_7^{2-} \; ion \; solution, \; then$$
$$x \; ml \times 1.8 \; mequiv/ml \; Cr_2O_7^{2-} = 9.6 \; mequiv \; Cr_2O_7^{2-}$$

whence

$x = 5.3$ ml of solution of $Cr_2O_7^{2-}$ ion for requirements in matching.

Collecting our several results for (a), (b), and (c), the required volumes of the solution of $Cr_2O_7^{2-}$ ion are, respectively, 1.6 ml, 2.8 ml, and 5.3 ml.

■ PROBLEM 15: To derive the initial concentration of a reagent from reaction product(s) obtained with it.

Addition of excess $Sr(NO_3)_2$ solution to 0.032 liter of aqueous K_2CO_3 solution of specific gravity 1.072 yields a total weight of 2.93 g of highly insoluble, precipitated $SrCO_3$. From this information, calculate

(a) the concentration formality and the normality, with respect to its specific employment here, of the K_2CO_3 solution. Likewise calculate the molarity of each of this solute's dissolved ions.

(b) rhe percentage by weight of dissolved K_2CO_3 in its prepared solutioh.

Solution. Reaction occurs by

$$Sr^{2+} + CO_3^{2-} \rightarrow SrCO_{3\,(s)}.$$

The stoichiometric significance of the equation is that the total number of moles of $SrCO_3$ precipitated by the Sr^{2+} ion (added in excess) must be equal to the total number of moles of CO_3^{2-} ion provided by the given volume of K_2CO_3 solution. As the gram-formula weight or mole of $SrCO_3$ is 147.63 g, we have

$$2.93 \text{ g SrCO}_3 \times \frac{1 \text{ mole SrCO}_3}{147.63 \text{ g SrCO}_3} = 0.020 \text{ mole SrCO}_3.$$

Hence, the K_2CO_3 solution must provide 0.025 mole of CO_3^{2-} ion which, in itself, must emanate from 0.020 g-formula weight of the K_2CO_3, in conformity with the stoichiometry of solubilization,

$$K_2CO_3 \rightarrow 2K^+ + CO_3^{2-}.$$

Clearly, the formality of the K_2CO_3 solution is numerically identical to the molarity of the CO_3^{2-} ion. This is to be equated as follows:

$$\text{total moles} = \text{liters} \times \text{molarity}$$

wherein substitution yields

$$\frac{0.020 \text{ mole}}{0.032 \text{ liter}} = 0.63 \text{ mole/liter}$$

$$= 0.63F \text{ K}_2\text{CO}_3 = 0.63M \text{ CO}_3^{2-}.$$

The concentration of K^+ ion in the prepared solution is

$$(2 \times 0.63M) \text{ or } 1.26M \text{ K}^+.$$

The metathetical reaction in which the solute is involved leads to the interpretation that

$$1 \text{ mole of K}_2\text{CO}_3 = 2 \text{ equivalents of K}_2\text{CO}_3,$$

consequently,

$$1F \text{ K}_2\text{CO}_3 = 2N \text{ K}_2\text{CO}_3$$

and

$$0.63F \text{ K}_2\text{CO}_3 = 1.26N \text{ K}_2\text{CO}_3.$$

Now, a 1.26 N solution of K_2CO_3 is, necessarily, $1.26N$ in K^+ ion and $1.26N$ in CO_3^{2-} ion. Collecting our several answers,

$$K_2CO_3 \text{ solution} = 0.63F \text{ and } 1.26N;$$

$$K^+ \text{ ion} = 1.26M \text{ and } 1.26N;$$

$$CO_3^{2-} \text{ ion} = 0.63M \text{ and } 1.26N.$$

(b) The percentage by weight of K_2CO_3 in the solution is derived from the relationship

$$\frac{\text{grams of solute}}{\text{grams of solution } (= \text{solute} + \text{solvent})} \times 100 = \text{weight } \% \text{ of solute.}$$

The stated specific gravity, in signifying the weight of one milliliter of solution, affords the following evaluation:

$$0.032 \text{ liter} \times \frac{1000 \text{ ml}}{1 \text{ liter}} \times \frac{1.072 \text{ g}}{1 \text{ ml}} = 34.30 \text{ g of total solution.}$$

We have already determined that the total solution provides 0.020 mole of K_2CO_3 solute. As the gram-formula weight of K_2CO_3 is 138.21 g, we will have

$$0.020 \text{ mole} \times \frac{138.21 \text{ g}}{1 \text{ mole}} = 2.76 \text{ g of pure } K_2CO_3 \text{ solute.}$$

Therefore

$$\text{weight}\% \text{ of solute} = \frac{2.76 \text{ g } K_2CO_3}{34.30 \text{ g solution}} \times 100$$

$$= 8.0\% \; K_2CO_3.$$

■ PROBLEM 16: To determine the volumes of two component solutions required to produce a stated concentration of a particular species common to both.

Calculate

(a) the volumes of separate solutions of $0.50F$ K_2SO_4 and $1.5F$ K_3PO_4, which, when mixed, without introduction of additional H_2O, will prepare 80.0 ml of a common solution that is $1.3N$ with respect to K^+ ion. Assume that the mixed volumes are exactly additive.

(b) the normalities of SO_4^{2-} and PO_4^{3-} ions in the prepared mixture. Ignore effects of hydrolysis.

Solution. We seek a concentration of K^+ ion expressed in terms of normality. We therefore, convert all other given concentrations to terms of K^+ normality and use the relationship

$$\underbrace{\frac{\text{total mequiv } K^+}{\text{ml} \times N}} = \underbrace{\frac{\text{total mequiv } K^+}{\text{ml} \times N}}.$$

We have

$$0.50F \ K_2SO_4 \rightarrow 1.0M \ K^+ + 0.50M \ SO_4^{2-}$$
$$\rightarrow 1.0N \ K^+ + 1.0N \ SO_4^{2-}$$

and

$$1.5F \ K_3PO_4 \rightarrow 4.5M \ K^+ + 1.5M \ PO_4^{3-}$$
$$\rightarrow 4.5N \ K^+ + 4.5N \ PO_4^{3-}.$$

If $x =$ milliliters of the K_2SO_4 solution to be taken, inasmuch as we desire a total volume of 80.0 ml of over-all solution,

$80.0 - x =$ milliliters of the K_3PO_4 solution required for mixing.

The arithmetical substitutions that fulfill the numerical identity of gram-equivalents required on left and right sides of the stated equality are then

$$\frac{K_2SO_4 \ \text{solution}}{x \ \text{ml} \times 1.0N \ K^+} + \frac{K_3PO_4 \ \text{solution}}{(80.0 - x) \ \text{ml} \times 4.5N \ K^+} = \frac{K_2SO_4 + K_3PO_4 \ \text{mixture}}{80.0 \ \text{ml} \times 1.3N \ K^+}$$

whence

$$3.5x = 256$$
$$x = 73.1 \ \text{ml of } K_2SO_4 \ \text{solution required for mixing}$$

and consequently,

$80.0 - 73.1 = 6.9$ ml of K_3PO_4 solution likewise required for mixing.

(b) It has been determined in the foregoing that in the individually separate K_2SO_4 solution the concentration of the SO_4^{2-} ion is $1.0N$. Inasmuch as in the mixing process 73.1 ml of this solution were raised to 80.0 ml, it follows that the concentration of the SO_4^{2-} ion will have been diminished to

$$\frac{73.1 \ \text{ml}}{80.0 \ \text{ml}} \times 1.0N \ SO_4^{2-} = 0.91N \ SO_4^{2-}, \ \textit{in mixture.}$$

Likewise, with respect to the $4.5N \ PO_4^{3-}$ established for its individually separate K_3PO_4 solution, the mixing dilution of the 6.9 ml taken results in lowering its concentration to

$$\frac{6.9 \ \text{ml}}{80.0 \ \text{ml}} \times 4.5N \ PO_4^{3-} = 0.39N \ PO_4^{3-}, \ \textit{in mixture.}$$

EXERCISES

1. In every instance calculate the concentration formality of the aqueous solution with respect to the pure solute, and the molarity of each of the ions that it supplies. Regard all solutes as completely ionic.

(a) 14.6 g of $CaCl_2 \cdot 6H_2O$ dissolved in 125.0 of solution.

(b) 20.6 mg of $MgSO_4 \cdot 7H_2O$ dissolved in 0.025 liter of solution.

(c) $Ca(NO_3)_2$ solution containing 12 mg to milliliter of Ca^{2+} ion.

(d) Na_2SO_4 solution containing 0.030 mole of Na^+ ion to 0.250 liter.

(e) HCl solution of specific gravity 1.19, containing 37.9 % by weight of dissolved HCl gas.

(f) H_2SO_4 solution of specific gravity 1.09, and containing 13.0 % by weight of anhydrous H_2SO_4. Consider the solute 100% dissociated.

(g) HNO_3 solution of specific gravity 1.11, containing 19.0% by weight of anhydrous HNO_3.

(h) 3.60 g of commercial grade KOH of 12.0% by weight of H_2O, dissolved in 50.0 ml of solution.

2. In each instance, calculate the weight in grams of the completely ionic solute required to prepare the stated solution.

(a) 75.0 ml of $0.40F$ KCl solution.

(b) 60.0 ml of K_2SO_4 solution of strength 15.0 mg to milliliter of K^+ ion.

(c) 80.0 ml of the double salt $KAl(SO_4)_2.12H_2O$, to yield a concentration of SO_4^{2-} ion of 0.150 mole/liter.

(d) 150.0 ml of the double salt $(NH_4)Al(SO_4)_2.12H_2O$, to yield a concentration of $0.250M$ of Al^{3+} ion.

(e) 125 ml of $3.00F$ NaOH, when the solid available is of a commercial grade containing 10.0% by weight of H_2O.

(f) 200.0 ml of $Ba(NO_3)_2$ solution to yield $0.500M$ NO_3^- The solid salt available for the purpose is the hydrate, $Ba(NO_3)_2.H_2O$.

(g) 250 ml of acidified $K_2Cr_2O_7$ solution containing minimum of solute required to oxidize fully the Fe^{2+} ion in an equal volume of acidified $0.80M$ Fe^{2+} ion. Redox products of reaction for which the solution is being prepared are Cr^{3+} and Fe^{3+}.

3. In each instance, calculate the maximum volume in milliliters of the solution sought that can be prepared from the given quantity of reagent when it is appropriately solubilized and/or diluted with H_2O.

(a) *Given*: 40.0 g of solid $CuSO_4.5H_2O$; *sought*: a solution of strength 15.0 mg of Cu^{2+} ion per milliliter.

(b) *Given*: 25.0 g of solid K_2SO_4; *sought*: a $0.300F$ K_2SO_4 solution.

(c) *Given*: 30.0 ml of $0.100F$ $Sr(NO_3)_2$ solution; *sought* a solution that is $0.0500M$ in NO_3^- ion.

(d) *Given*: 13.5 ml of aqueous NH_3 solution of specific gravity 0.900 and containing 28.0% by weight of dissolved NH_3 gas; *sought*: a $7.50F$ NH_3 solution.

(e) *Given:* 15.0 ml of $17.0F$ $HC_2H_3O_2$ solution; *sought:* $HC_2H_3O_2$ solution of specific gravity 1.05, containing 35.0% by weight of anyhdrous acetic acid.

(f) *Given:* 10.0 ml of $CuSO_4$ solution containing 20.0 mg of Cu^{2+} ion per milliliter; *sought:* a solution of concentration 15.0 mg Cu^{2+} ion per milliliter.

(g) *Given:* 35.0 ml of $0.600N$ $BaCl_2$ solution; *sought:* a solution that is $0.200F$ with respect to $BaCl_2$.

(h) *Given:* 25.0 ml of solution containing a total dissolved weight of 1.60 g of $Ca(NO_3)_2$; *sought:* a solution that is $0.0150N$ in NO_3^- ion, for use in nonredox reaction.

(i) *Given:* 60.0 ml of a solution of $Al_2(SO_4)_3$ containing 2.00 mequiv of $Al_2(SO_4)_3$ per milliliter; *sought:* a solution that is $0.800M$ with respect to Al^{3+} ion.

4. In each instance, calculate with respect to the stated quantity of ion the number of millimoles and the number of milliequivalents represented thereof; also the maximum number of milliliters of $0.500N$ aqueous solution that it will supply for its specified use.

(a) 1.50 g of Cr^{3+} ion, for reaction conversion to $Cr(OH)_3$.

(b) 1.50 g of Cr^{3+} ion, for reaction conversion to $Cr_2O_7^{2-}$ ion.

(c) 1.50 g of $Cr_2O_7^{2-}$ ion, for reaction conversion to Cr^{3+} ion.

(d) 2.50 g of $Fe(CN)_6^{4-}$ ion, for reaction conversion to $Ag_4[Fe(CN)_6]$.

(e) 2.50 g of $Fe(CN)_6^{4-}$ ion, for reaction conversion to $Fe(CN)_6^{3-}$.

(f) 0.500 g of S^{2-} ion, for reaction conversion to CdS.

(g) 0.500 g of S^{2-} ion, for reaction conversion to elementary S^0.

(h) 0.300 g of Mn^{2+} ion, for reaction conversion to MnO_2.

(i) 0.300 g of Mn^{2+} ion, for reaction conversion to Mn^{3+} ion.

(j) 0.300 g of Mn^{2+} ion, for reaction conversion to MnO_4^- ion.

(k) 0.100 g of $Co(NH_3)_6^{2+}$ ion, for reaction conversion to Co^{2+} ion.

(l) 0.100 g of $Co(NH_3)_6^{2+}$ ion, for reaction conversion to $Co(NH_3)_6^{3+}$ ion.

5. Calculate the number of grams, the number of moles, and the number of gram-equivalents of Cu^{2+} ion precipitated as CuS by the total quantity of H_2S formed when 1.2 g of FeS react completely with excess HCl solution by

$$FeS + 2H^+ \rightarrow Fe^{2+} + H_2S.$$

6. Calculate the volume in milliliters of $2.5F$ NH_3 aqueous ($2.5F$ NH_4OH) required for the quantitatively complete precipitation of the Fe^{3+} ion, as $Fe(OH)_3$, from a solution containing a total dissolved weight of 0.18 g of $Fe(NO_3)_3 \cdot 9H_2O$.

7. Calculate the weight in grams of Hg_2^{2+} ion that will be precipitated as the highly insoluble Hg_2Cl_2 by

(a) 5.0 ml of $3.0F$ HCl solution.

(b) 5.0 mequiv of Cl^- ion.

(c) 5.0 mequiv of $CaCl_2$.

(d) 0.50 g of NaCl.

(e) 0.50 g of $BaCl_2$.

(f) 0.050 liter of $0.40N$ $BaCl_2$ solution.

8. When 13.5 g of a certain solid mixture is fully water-solubilized, the $Al_2(SO_4)_3$ component therein is found to react with quantitative completeness with an excess of a solution of Pb^{2+} ion, precipitating 0.0480 g-equiv of $PbSO_4$. The other components of the solubilized mixture are found to be chemically unreactive toward the Pb^{2+} ion. Calculate from this information the percent by weight of $Al_2(SO_4)_3$ in the mixture.

9. Calculate the number of milliliters of $3.5M$ solution of NO_3^- required to react fully in acidic medium with each of the following:

   (a) 50 mequiv of PbS; reaction products Pb^{2+}, NO, S^0.

   (b) 50 of $0.65F$ ml solution of H_3AsO_3; reaction products NO, H_3AsO_4.

   (c) 50 of $2.0F$ solution of H_3AsO_3; reaction products, NO_2, H_3AsO_4.

   (d) 0.50 g of metallic Mg^0; reaction products, Mg^{2+}, NH_4^+.

   (e) 0.50 g of Al^{3+} ion; reaction products Al^{3+}, NO.

10. Given 15.0 ml of 0.30 % Fe^{2+} ion solution, to be used in its entirety for reduction of $KMnO_4$ in acidic medium, redox products being Fe^{3+} and Mn^{2+}, calculate the quantity of $KMnO_4$ reduced, in terms of each of the following:

   (a) milliequivalents.

   (b) grams.

   (c) millimoles.

   (d) milliliters of $0.30F$ solution of $KMnO_4$.

11. Given 30.0 ml of a solution $0.500M$ in $Cr_2O_7^{2-}$ ion, to be used in its entirety for oxidation of KI in acidic medium, redox products being Cr^{3+} and I_2, calculate the maximum volume, in milliliters, of $0.400F$ solution of KI that is needed for complete reduction of the total available supply of dichromate ion.

12. Calculate the weight in grams of H_2O_2 that undergoes reduction to H_2O in its oxidation of 0.30 g of Cr^{3+} ion to $Cr_2O_7^{2-}$ ion in acidic medium.

13. Calculate the weight in grams of NO_3^- ion that is dissolved in 180 ml of $0.35N$ solution of $NaNO_3$ prepared for use in an alkaline medium wherein it undergoes reduction to NO_2^- ion.

14. How many milliliters of $1.2N$ solution of Na_2SO_3 can be prepared from 5.0g of solid Na_2SO_3, if the solution is to be used as a reductant in alkaline medium wherein the anion in the solute undergoes oxidation to SO^{2-} ion?

15. A total weight of 0.054 g of the compound $HgCl_2$ was found to be reduced to Hg_2Cl_2 when treated with 12.0 ml of $SnCl_2$ solution, the latter being completely used up in the reaction. Calculate from this information the normality of the $SnCl_2$ solution.

16. A total weight of 0.054 g of Hg^{2+} ion was found to be reduced to Hg_2Cl_2 when treated with 12.0 ml of $SnCl_2$ solution, the latter being completely used up in the reaction. Calculate the normality of the $SnCl_2$ solution.

17. The reaction between $Cr_2O_7^{2-}$ and I^- ions in acidic medium yields Cr^{3+} ion and I_2. Calculate the quantity of each of the following required for such reaction with 50.0 ml of $0.0300M$ solution of I^- ion:

   (a) millimoles of $Cr_2O_7^{2-}$ ion.

   (b) grams of $Cr_2O_7^{2-}$ ion.

   (c) grams of $K_2Cr_2O_7$.

   (d) milliequivalents of $K_2Cr_2O_7$.

   (e) milliliters of $0.200M$ $Cr_2O_7^{2-}$ ion.

   (f) milliliters of $0.200F$ $K_2Cr_2O_7$.

   (g) milliliters of $0.200N$ $K_2Cr_2O_7$.

   (h) milliliters of a $K_2Cr_2O_7$ solution that is $0.200M$ in K^+ ion.

18. Calculate which of the following two oxidants in alkaline medium oxidizes the greater quantity of $Fe(OH)_2$ to $Fe(OH)_3$, stating the quantity of the comparative amount of excess of $Fe(OH)_2$ so oxidized in terms of both milliequivalents and grams: 15.0 m of 0.85F H_2O_2 (reduced to OH^-ion), or 15.0 ml of 1.50F $KMnO_4$ (reduced to MnO_2).

19. An 8.00F solution of H_2SO_4 has a density of 1.440 g/ml. Calculate the percent by weight of anhydrous solute H_2SO_4 therein.

20. A 12.0F solution of HCl contains 37.8% by weight of gaseous anhydrous solute· Calculate the density of the solution in grams/milliliter.

21. A solution of NaOH has a specific gravity of 1.22 and contains 19.7% by weight of pure anhydrous solute. Calculate the concentration formality of the solution.

22. A quantitative (virtually complete) precipitation of AgCl ensues when 40.0 ml of a solution containing a dissolved total weight of 2.50 g of $BaCl_2$ are mixed with 50.0 ml of a solution containing a dissolved weight of 0.750 g of $AgNO_3$ to the milliliter. Calculate
   (a) the number of gram-formula weights of the insoluble AgCl that will be formed.
   (b) which of the reacting ions is in excess, stating the quantities of its total excess in terms of moles, of grams, and of gram-equivalents.
   (c) the individual concentrations, in moles/liter, of Ba^{2+} and NO_3^- ions in the mixture. Assume volumes of the given solutions to be precisely additive when mixed.

23. Given a solution of H_2SO_4 of density 1.83 g/ml and containing 98.0% by weight of anhydrous solute. Calculate the solume in milliliters, of this solution, required to yield 250 ml of 9.00F H_2SO_4 when diluted with water.

24. Calculate the volume, in milliliters, of aqueous NH_3 solution, of specific gravity 0.900 and containing 28.0% of anhydrous solute, required to prepare by dilution 60.0 of a 3.00F solution of NH_3.

25. It requires 6.50 ml of a given solution of a $KMnO_4$ to oxidize 0.0650 g of I^-ion in acidic medium. If redox products are Mn^{2+} ion and I_2, what is
   (a) the normality of the $KMnO_4$ solution?
   (b) the formality of the $KMnO_4$ solution?
   (c) the number of gram-equivalents of I^- ion oxidized?
   (d) the number of milliequivalents of MnO^- ion reduced?

26. Calculate the volume in milliliters of 0.0600F solution of $AgNO_3$ required to precipitate the CrO_4^{2-} ion (as Ag_2CrO_4) in 45.0 ml of a 0.600F solution of K_2CrO_4.

27. Calculate the weight of H_2S in grams which in acidic reaction with 5.6 g of dissolved $KMnO_4$ produces MnO_2 (precipitated) and SO_4^{2-} ion as redox products.

28. To 25 ml of a 0.30F solution of H_2SO_4 are added 100 ml of a 0.50F solution of H_2SO_4. Assuming volumes to be exactly additive, calculate the concentration formality of the H_2SO_4 of the final mixture.

29. Calculate the volume in milliliters of $0.150F$ NaOH required in each of the following instances to react with 25.0 ml of a $0.200\ F$ solution of H_3PO_4, reaction conforming to

(a) $OH^- + H_3PO_4 \rightarrow H_2O + H_2PO_4^-$.

(b) $2OH^- + H_3PO_4 \rightarrow 2H_2O + HPO_4^{2-}$.

(c) $3OH^- + H_3PO_4 \rightarrow 3H_2O + PO_4^{3-}$.

30. The following solutions are mixed to provide complete homogeneity: 30.0 ml of $0.300F$ KNO_3, 20.0 ml of $0.200F$ K_2SO_4, and 75.0 ml of $0.200F$ HNO_3. Regard the volumes of the component solutions to be precisely additive. Calculate, with respect to the final mixture,

(a) the concentrations, in moles per liter, of each of the ions present therein.

(b) the normality of the SO_4^{2-} ion when the mixture is to be used for the metathetical precipitation of $BaSO_4$.

(c) the normality of the NO_3^- ion when the mixture is to be used in a redox reaction in which the NO_3^- ion is converted exclusively to NO gas.

(d) the weight in grams of pure NaOH that will be required to neutralize the total volume of the prepared mixture when it is solubilized therein.

31. A solution is prepared by mixing thoroughly 25.0 ml each of the following solutions: $0.0200F$ $K_2Cr_2O_7$, $0.840F$ KI, and $0.360F$ H_2SO_4 (consider the last to be 100% ionized to H^+ and SO_4^{2-} ions). The admixed volumes are to be regarded as precisely additive. If the redox reaction that occurs produces Cr^{3+} ion and I_2, what are

(a) the number of gram-equivalents of I_2 formed?

(b) the number of moles of Cr^{3+} ion formed?

(c) the final concentration molarities of I^-, SO_4^{2-}, and Cr^{3+} ions in the mixture at equilibrium?

32. In the titration of a 12.4-ml sample of commercial vinegar of density 1.01 g/ml with 0.20 NaOH solution, it is found that the equivalence point of the titration is reached when 49.6 ml of the base have been introduced. Calculate therefrom the percent by weight of acetic acid in the vinegar.

33. The concentration of a given solution of oxalic acid has been calculated at $0.0650N$ for its reaction in conformity with

$$H_2C_2O_4 + OH^- \rightarrow H_2O + HC_2O_4^-.$$

Calculate

(a) the weight of $H_2C_2O_4$, in grams to 100 ml, provided by the given solution if it is to conform to the stated neutralization reaction.

(b) the normality of the same given solution were it to be used instead for neutralization by

$$H_2C_2O_4 + 2OH^- \rightarrow 2H_2O + C_2O_4^{2-}.$$

(c) The numbers of milliequivalents and of grams, respectively, of $Ca(OH)_2$ that are required for reaction with 125-ml samples of the given $H_2C_2O_4$ solution separately in accordance with the equations:

     (1) $H_2C_2O_4 + Ca(OH)_2 \rightarrow HC_2O_4^- + Ca(OH)^+ + H_2O$

     (2) $H_2C_2O_4 + Ca(OH)_2 \rightarrow CaC_2O_4 + 2H_2O$.

34. When 25.0 ml of a certain HCl solution were diluted with water to 80.0 ml, the final solution was found to contain 0.0720 mequiv to the milliliter of solute HCl. Calculate both the concentration normality (metathetical) and the formality of the initial undiluted solution.

35. One hundred milliliters of an aqueous $6.00F$ solution of KOH were prepared, using commercial grade solute containing 12.0% by weight of H_2O. Calculate therefrom the weight in grams of this commercial solute that was dissolved in the preparation.

36. A solution of K_2CrO_4 was added in excess to 50.0 ml of a certain solution of $Pb(NO_3)_2$, with consequent quantitative precipitation of the cation of the latter in the form of $PbCrO_4$. After being washed and dried, the precipitate was found to weigh 2.4355 g. Calculate, therefrom,
   (a) the formality and the normality of the $Pb(NO_3)_2$ solution.
   (b) the molarity and the normality of both Pb^{2+} and NO_3^- ions in the solution at equilibrium.
   (c) the number of milliequivalents and the number of millimoles of CrO_4^{2-} ion that were simultaneously precipitated.

37. Calculate the number of grams of solid K_2SO_4 (completely ionic) required to prepare 100 ml of each of the following solution. Where necessary, assume the density of H_2O to be 1.0 g/ml.
   (a) $0.600F\ K_2SO_4$.
   (b) $0.600M\ K^+$ ion.
   (c) $0.600N\ SO_4^{2-}$ ion.
   (d) $0.600m$ (molal) K^+ ion.
   (e) $0.600m$ (molal) in total combined concentrations of K^+ and SO_4^{2-} ions.
   (f) 0.600 g-equiv of SO_4^{2-} ion in total.
   (g) 0.600 mequiv per ml of K^+ ion.

38. Calculate the number of milliliters, respectively, of a $2.00F$ solution of $CaCl_2$ and a $1.60F$ solution HCl that must be mixed, without additional H_2O, in order to yield 98.0 ml of a common solution that is $2.00N$ in Cl^- ion. Regard the admixed volumes to be precisely additive.

39. Calculate the volume in milliliters of aqueous NH_3 solution, of specific gravity 0.980 and assaying 5.10% by weight of anhydrous solute, that contains one gram-formula weight of the latter.

40. Calculate the weight in grams of H_2O that will be needed to solubilize sufficient $MgCrO_4.7H_2O$ to produce 100g of a solution that is 12.0% by weight of $MgCrO_4$ (unhydrated).

41. Calculate the total number of milliliters to which 60.0 ml of a $8.00F$ solution of H_2SO_4 must be diluted with water in order to yield a solution of density 1.09 g/ml and containing 13.0% by weight of anhydrous solute.

42. Calculate the volumes in milliliters of an $18.0F$ solution of H_2SO_4 required for the following separate fulfillments:
   (a) the metathetical conversion of 50.0 mequiv of Ba^{2+} ion to precipitated $BaSO_4$.

(b) the metathetical conversion of 50.0 mequiv of M^{2+} ion (hypothetical) to $M(HSO_4)_2$.

(c) the oxidation of 50.0 g of Br^- ion to elemental Br_2 and the concomitant reduction of H_2SO_4 to SO_2.

(d) the oxidation of 50.0 moles of Q^{2-} ion (hypothetical) to elemental Q_2^0 and the concomitant reduction of H_2SO_4 to H_2S.

43. At $t°C$, it is found that 54.0 ml of a given H_2O_2 solution, of density $= 1.240$ g/ml, were just sufficient to react with 24.0 ml of $0.300F$ $K_2Cr_2O_7$ in acidic medium. Redox products are Cr^{+3} ion and O_2 gas. Calculate, therefrom, the percent by weight of anhydrous H_2O_2 present in its separate given solution.

44. Calculate the weight in grams of Cl_2 that in acidic medium oxidizes the same quantity of HNO_2 to NO_3^- ion as does

(a) 0.30 liter of $0.12F$ $KMnO_4$.

(b) 0.30 g of $KMnO_4$.

(c) 0.30 g of MnO_4^- ion.

(d) 0.30 mole of MnO_4^- ion.

(e) 0.30 g-formula weight of $KMnO_4$.

(f) 0.30 mequiv of MnO_4^- ion.

The respective products of reduction are in all instances Cl^- ion and Mn^{2+} ion.

45. Calculate the respective minimum normalities and minimum molarities of different solution of Mn^{2+} ion, each of 125.0 ml-volume, which, when used in their entireties, will serve the following requirements:

(a) The metathetical precipitation, in basic medium, of 500 mg of $Mn(OH)_2$.

(b) The metathetical precipitation, in basic medium, of 500 mg of Mn^{2+} ion from solution.

(c) The redox precipitation of 500 mg of MnO_2.

(d) The redox conversion, in acidic medium, to 500 mg of MnO_4^- ion

(e) The equality of capacity for reduction, in acidic medium, with that of 125.0 ml of $0.250\ M\ Q^-$ ion (hydothetical) when both Mn^{2+} and Q^- undergo oxidation, respectively, to MnO_4^- and Q^{2+}.

(f) The equality of capacity for reduction, in basic medium, with 60.0 ml of $0.450M\ X_3^{2-}$ ion (hypothetical) when both Mn^{2+} and X_3^{2-} undergo oxidation, respectively, to MnO_3^{2-} and X_2^-.

46. Calculate the volume, in milliliters, of a $0.25F$ solution of $K_2Cr_2O_7$ that oxidize the identical quantity of Sn^{2+} ion in acidic medium (redox products, Cr^{3+} and Sn^{IV} — that is, tin in an oxidation state of $+4$) as

(a) 40 ml of a $0.18F$ solution of $KMnO_4$ (redox products, Sn^{IV} and Mn^{2+}).

(b) 30 mmoles of I_3^- ion (redox products, Sn^{IV} and I^-).

(c) 2.5 g of NO_3^- ion (redox products, Sn^{IV} and NO).

47. A 0.372-g sample of commercial KOH containing inert material is dissolved in water and titrated with a $0.120F$ solution of HCl. It is found that 44.5 ml of this acid are required to reach the equivalence point of the titration. Calculate, therefrom,

(a) the percent purity of KOH in the given sample.

(b) the volume of solution, in milliliters, in which the given weight of the commercial sample would have to be dissolved in order to yield, before titration, a concentration of OH^- ion of 0.0800 mole/liter.

48. Calculate the number of grams of MnO_4^- ion required to react, in acidic medium, with

(a) 0.460 g-equiv of H_2S, to yield SO_4^{2-} ion and MnO_2.
(b) 0.460 g of H_2S, to yield SO_4^{2-} ion and MnO_2.
(c) 0.460 g-equiv of H_2S, to yield elemental S^0 and MnO_2.
(d) 0.460 g-equiv of $H_2C_2O_4$, to yield Mn^{2+} ion and CO_2.
(e) 0.460 g of $H_2C_2O_4$, to yield Mn^{2+} ion and CO_2.

49. Calculate for the reaction,

$$H_3PO_4 + 2HBO_3^{2-} \rightarrow HPO_4^{2-} + 2H_2BO_3^-$$

(a) the number of milliequivalents present in a millimole of H_3PO_4.
(b) the weight of HBO_3^{2-} ion, in grams, that will have reacted with 20.0 ml of $0.0300F$ H_3PO_4 by the stoichiometry of the written equation.

50. A given solution of K_2CrO_4 is calculated to be $0.50N$ in CrO_4^{2-} ion when used as a reagent for the metathetical precipitation of Ba^{2+} ion (as $BaCrO_4$). Calculate the normality of this same solution when used, respectively,

(a) To react with aqueous HNO_3 solution to form $Cr_2O_7^{2-}$ ion.
(b) To undergo reduction in alkaline medium to $Cr(OH)_4^-$ ion.
(c) To supply the chromium content for the formation of CrO_5 (perchromic anhydride) using for the purpose of conversion H_2O_2 in HNO_3 medium. Given: One of the oxygen atoms in the CrO_5 is of -2 oxidation state; the other four are peroxy atoms of -1 oxidation state.

51. Given the following indicators and their respective midpoints of change of color:

Indicator	L	M	N	P	Q	R	S	T	W	X	Y	Z
pH at midpoint	3.9	4.5	5.1	6.7	7.0	7.3	8.0	8.8	9.4	9.7	10.0	10.5

By calculation, determine which of these indicators would, in each instance, have properly identified the equivalence points of the following titrations:

(a) 30.0 ml of $0.10F$ $HC_2H_3O_2$, with titrant $0.25F$ NaOH.
(b) 22.0 ml of $0.30F$ NH_3 aqueous, with titrant $0.20F$ HCl.
(c) 70.0 ml of $0.15F$ NH_3 aqueous, with titrant $0.42F$ $HC_2H_3O_2$.
(d) 60.0 ml of $0.20F$ NaOH, with titrant $0.30F$ HCl.

52. The complete titration of a solution containing 0.662 g of a dissolved acid (identity concealed) requires 42.5 ml of $0.200F$ KOH. Calculate therefrom the weight, in grams, of one gram-equivalent of this acid.

53. Two samples of the same prepared solution of KOH, identical in volume, are separately titrated, one with a $0.050F$ solution of HCl the other with $0.040F$ $HC_2H_3O_2$. If 37.6 ml of HCl solution were used to attain the equivalence point of its respective titration, what will be the volume, in milliliters, of the $HC_2H_3O_2$ solution that likewise is required to reach the equivalence point of the sample with which it is in reaction?

APPENDIX A: ANSWERS TO EXERCISES

CHAPTER 2: The Nature of Electrolytes.

1. (a)

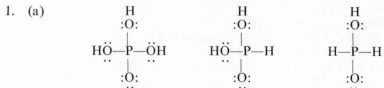

orthophosphoric acid   phosphorus acid   hypophosphorus acid

(b)

$$
\begin{array}{cc}
\text{H} & \text{H} \\
\text{:O:} & \text{:O:} \\
| & | \\
\text{HÖ--P--ÖH} & \text{HÖ--P--H} \\
\text{(triprotic)} & \\
& \text{:O:} \\
& \text{(diprotic)}
\end{array}
$$

(c)   (i) the answer is NO; $H_2PO_2^-$ is not an oxyanion (no hydroxy groups are present).
      (ii) the answer is YES; $H_2PO_3^-$ is an oxyanion (one hydroxy group is present).

2. $NaOH < Mg(OH)_2 < Al(OH)_3 < Si(OH)_4 (= H_4SiO_4) < H_3PO_4 < H_2SO_4 < HClO_4$.

3. (a) $H_3PO_4 > H_3PO_3 > H_3PO_2$.
   (b) $HClO_4 > HClO_3 > HClO_2 > HClO$.
   (c) $(HO)ClO_3; (HO)_2SO_2; (HO)_2SO; (HO)_3As$.
   (d)   A lone oxygen atom acquires a greater electron density than one that is forced to share charge distribution with a hydrogen atom. Hence, lone oxygen atoms cause a shift to themselves of the bonding electron-pairs in the H–O linkages. The hydrogen atom is thus at a disadvantage with respect to requisite charge atmosphere in its bond and, consequently, ionizes more readily to seek greater stability with a more effective electron-pair donor.

4. $As(OH)_3 < Sb(OH)_3 < Bi(OH)_3$

5. (a) $Te^{2-} < Se^{2-} < S^{2-} < O^{2-}$.
   (b) $AsH_3 < PH_3 < NH_3$.
   (c) $HIO_4 < HBrO_4 < HClO_4$.
   (d) $HO^- > HS^- > HSe^- > HTe^-$.
   (e) $H_2SO_4 < H_2SeO_4 < H_2TeO_4$.
   (f) $F^- > Cl^- > Br^- > I^-$.
   (g) $HF < HCl < HBr < HI$.
   (h) $HF > HCl > HBr > HI$.

(i) $\quad$ HF > HCl > HBr > HI.

(j) $\quad$ HI < HBr < HCl < HF.

6. (a) $\quad$ anhydrous: $\quad HF + HF \rightarrow H_2F^+ + F^-$

$\qquad\qquad\qquad$ acid$_1$ $\quad$ base$_2$ $\quad$ acid$_2$ $\quad$ base$_1$

$\qquad$ aqueous: $\quad HF + H_2O \rightarrow H_3O^+ + F^-$

$\qquad\qquad\qquad$ acid$_1$ $\quad$ base$_2$ $\quad$ acid$_2$ $\quad$ base$_1$

(b) $\quad$ HF

7.

$$\underset{H_4P_2O_7}{HO-\overset{\overset{H}{\overset{\displaystyle |}{:O:}}}{\underset{\underset{:O:}{\displaystyle |}}{P}}-\ddot{O}-\overset{\overset{H}{\overset{\displaystyle |}{:O:}}}{\underset{\underset{:O:}{\displaystyle |}}{P}}-\ddot{O}H} \qquad\qquad \underset{H_5P_3O_{10}}{HO-\overset{\overset{H}{\overset{\displaystyle |}{:O:}}}{\underset{\underset{:O:}{\displaystyle |}}{P}}-\ddot{O}-\overset{\overset{H}{\overset{\displaystyle |}{:O:}}}{\underset{\underset{:O:}{\displaystyle |}}{P}}-\ddot{O}-\overset{\overset{H}{\overset{\displaystyle |}{:O:}}}{\underset{\underset{:O:}{\displaystyle |}}{P}}-\ddot{O}H}$$

8. (a)

$$\left[\cdot B:\right]^0 + 3\left[\cdot\ddot{F}:\right]^0 \rightarrow \left[\overset{:\ddot{F}:}{\underset{}{:\ddot{F}-B-\ddot{F}:}}\right]^0, \text{ three normal covalent bonds}$$

$$\left[\overset{:\ddot{F}:}{:\ddot{F}-B-\ddot{F}:}\right]^0 + \left[:\ddot{F}:\right]^- \rightarrow \left[\overset{:\ddot{F}:}{\underset{:\ddot{F}:}{:\ddot{F}-B-\ddot{F}:}}\right]^-, \text{ three normal covalent and one coordinate covalent bonds, all equivalent.}$$

(b)

$$\left[Be:\right]^0 + 2\left[\cdot\ddot{F}:\right]^0 \rightarrow \left[Be\right]^{2+} + 2\left[:\ddot{F}:\right]^-, \text{ ionic bonds}$$

$$\left[Be\right]^{2+} + 4\left[:\ddot{F}:\right]^- \rightarrow \left[\overset{:\ddot{F}:}{\underset{:\ddot{F}:}{:\ddot{F}-Be-\ddot{F}:}}\right]^{2-}, \text{ four coordinate covalent bonds.}$$

(c)

$$\left[:Be\right]^0 + 2\left[\cdot\ddot{Cl}:\right]^0 \rightarrow \left[:\ddot{Cl}-Be-\ddot{Cl}:\right]^0, \text{ two normal covalent bonds.}$$

$$\left[:\ddot{Cl}-Be-\ddot{Cl}:\right]^0 + 2\left[:\ddot{Cl}:\right]^- \rightarrow \left[\overset{:\ddot{Cl}:}{\underset{:\ddot{Cl}:}{:\ddot{Cl}-Be-\ddot{Cl}:}}\right]^{2-}$$

two normal covalent and two coordinate covalent bonds, all equivalent.

9.

$$\left[\overset{H}{\underset{\ddot{H}}{H:\ddot{C}}}\right]^+, \textit{Carbonium ion}$$

$$\left[\overset{H}{\underset{\ddot{H}}{:\ddot{C}:H}}\right]^-, \textit{the carbanion}$$

10. $[Fe(H_2O)_6]^{3+} + H_2O \rightleftharpoons [Fe(H_2O)_5(OH)]^{2+} + H_3O^+$
$[Fe(H_2O)_5(OH)]^{2+} + H_2O \rightleftharpoons [Fe(H_2O)_4(OH)_2]^+ + H_3O^+$
$[Fe(H_2O)_4(OH)_2]^+ + H_2O \rightleftharpoons [Fe(H_2O)_3(OH)_3]^0 + H_3O^+$.

11. The answer is NO. The electronegativities of the hydrogen and phosphorus atoms are so very nearly alike (≈ 2.1) that the phosphorus-bonded hydrogens do not ionize. Consequently, they are not acidic — as they would be were they bonded instead to the highly electronegative oxygen atoms.

CHAPTER 4: Conductance of Ions.

1. (a) 4.81×10^{-4} mho;
    (b) $0.0680N$;
    (c) 195.0 mhos
2. 1.3%
3. (a) $0.0250N$ HX;
    (b) 1.01×10^{-2} mho;
    (c) 99.0 ohms;
    (d) 73.1%
4. $\Lambda_0 HC_2H_3O_2 = 387$ mhos
5. (a) 0.910 cm^{-1};
    (b) 152 ohms;
    (c) 0.00658 mho
6. 0.96 ohm
7. (a) 42.1 mhos;
    (b) 5.77%;
    (c) $1.14 \times 10^{-3}M$ H$^+$, $1.14 \times 10^{-3}M$ Y$^-$

CHAPTER 5: Colligative Properties of Ionic Solutions.

1. 5.38% ion-pairing
2. 20.2% ion-clustering
3. $0.0987F$ MgCl$_2$
4. (a) 0.690 m;

(b) $-1.28°C$;
(c) $100.36°C$
5. 7.5%
6. (a) $0.304m$;
    (b) 0.185 mole $MA_3/1000$ g H_2O, 0.0296 mole
            $M^{3+}/1000$ g H_2O, 0.0888 mole $A^-/1000$ g H_2O;
    (c) 13.8%
7. 2.62 g of K_2SO_4

CHAPTER 7: Balancing the Redox Equation.

1. (a) $4Hg_2CrO_4 \rightarrow 8Hg^0 + 2Cr_2O_3 + 5O_2$.
    (b) $4Co^{2+} + O_2 + 24NH_3 + 2H_2O \rightarrow 4Co(NH_3)_6^{3+} + 4OH^-$.
    (c) $K_2Cr_2O_7 + 4NaCl + 6H_2SO_4 \rightarrow 2CrO_2Cl_2 + 2KHSO_4$
            $+ 4NaHSO_4 + 3H_2O$.
    (d) $5NaHSO_3 + 2NaIO_3 \rightarrow 3NaHSO_4 + 2Na_2SO_4 + I_2 + H_2O$.
    (e) $2 Cr_2O_3 + 4K_2CO_3 + 3O_2 \rightarrow 4K_2CrO_4 + 4CO_2$.
    (f) $4Au + 8KCN + O_2 + 2H_2O \rightarrow 4KAu(CN)_2 + 4KOH$.
    (g) $CH_3OH + 6NaClO_3 + 3H_2SO_4 \rightarrow CO_2 + 6ClO_2$
            $+ 3Na_2SO_4 + 5H_2O$.
    (h) $8HI + H_2SO_4 \rightarrow 4I_2 + H_2S + 4H_2O$.
    (i) $2Ca_3(PO_4)_2 + 6SiO_2 + 10C \rightarrow 6CaSiO_3 + P_4 + 10CO$.
    (j) $2KClO_3 + 4HCl \rightarrow 2KCl + 2ClO_2 + Cl_2 + 2H_2O$.
    (k) $4Fe(CrO_2)_2 + 8Na_2CO_3 + 7O_2 \rightarrow 2Fe_2O_3 + 8Na_2CrO_4$
            $+ 8CO_2$.
    (l) $3MnO_2 + 3Na_2CO_3 + KClO_3 \rightarrow Na_2MnO_4 + KCl + 3CO_2$.
    (m) $Fe(CN)_6^{4-} + 10SO_4^{2-} + 6H_2O + 22H^+ \rightarrow Fe^{2+} + 10HSO_4^-$
            $+ 6NH_4^+ + 6CO$.
    (n) $TiCl_4 + 2Na_2S_2O_3 + H_2O \rightarrow H_2TiO_3 + 4NaCl + 2SO_2$
            $+ 2S^0$.
    (o) $3(NH_4)_2SO_4 \rightarrow N_2 + 4NH_3 + 3SO_2 + 6H_2O$.
2. (a) $P_4 + 3KOH + 3H_2O \rightarrow PH_3 + 3KH_2PO_2$.
    (b) $2KMnO_4 + 16HCl \rightarrow 2KCl + 2MnCl_2 + 5Cl_2 + 8H_2O$.
    (c) $As_2O_3 + 6Zn^0 + 12HCl \rightarrow 2As^0 + 6ZnCl_2 + 3H_2 + 3H_2O$.
    (d) $2Al^0 + NaH_2AsO_3 + NaOH + 4H_2O \rightarrow 2NaAl(OH)_4$
            $+ AsH_3$.
    (e) $3CdS + 8HNO_3 \rightarrow 3Cd(NO_3)_2 + 3S^0 + 2NO + 4H_2O$.
    (f) $2FeSO_4 + Na_2O_2 + 2NaOH + 2H_2O \rightarrow 2Fe(OH)_3$
            $+ 2Na_2SO_4$.

(g) $2F_2 + 2NaOH \rightarrow OF_2 + 2NaF + H_2O.$

(h) $5PbO_2 + 2Mn(NO_3)_2 + 4HNO_3 \rightarrow 4Pb(NO_3)_2 + Pb(MnO_4)_2 + 2H_2O.$

(i) $3U_3O_8 + 18HCl + 2HNO_3 \rightarrow 9(UO)_2Cl_2 + 2NO + 10H_2O.$

(j) $3FeS + 12HNO_3 \rightarrow 3Fe(NO_3)_3 + 3S^0 + 6H_2O + 3NO.$

(k) $Cu(NH_3)_4SO_4 + Na_2S_2O_4 + 2H_2O \rightarrow Cu^0 + 2(NH_4)_2SO_3 + Na_2SO_4.$

(l) $5Ti(SO_4)_3 + 2KMnO_4 + 8H_2SO_4 \rightarrow 10Ti(SO_4)_2 + 2MnSO_4 + K_2SO_4 + 8H_2O.$

(m) $4MnSO_4 + O_2 + 10H_2O + 8NH_3 \rightarrow 4Mn(OH)_3 + 4(NH_4)_2SO_4.$

3. (a) $6Co(NO_2)_6^{3-} + 2NH_4^+ + 8C_2H_3O_2^- \rightarrow 6Co^{2+} + 36NO_2^- + 8HC_2H_3O_2 + N_2.$

(b) $2Fe(OH)_3 + 3Cl_2 + 10OH^- \rightarrow 2FeO_4^{2-} + 6Cl^- + 8H_2O.$

(c) $2BiO(OH) + 3HSnO_2^- + 3OH^- + 5H_2O \rightarrow 2Bi^0 + 3Sn(OH)_6^{2-}.$

(d) $3C_2H_5OH + 2Cr_2O_7^{2-} + 16H^+ \rightarrow 3HC_2H_3O_2 + 4Cr^{3+} + 11H_2O.$

(e) $2SbO^+ + 3S_2O_3^{2-} + 4HC_2H_3O_2 \rightarrow Sb_2OS_2 + 4SO_2 + 4C_2H_3O_2^- + 2H_2O.$

(f) $P_4 + 3OH^- + 3H_2O \rightarrow PH_3 + 3H_2PO_2^-.$

(g) $IO_3^- + 2I_2 + 5Cl^- + 6H^+ \rightarrow 5ICl + 3H_2O.$

(h) $4CrO_5 + 12H^+ \rightarrow 4Cr^{3+} + 7O_2 + 6H_2O.$

(i) $5S_2O_8^{2-} + 2Mn^{2+} + 8H_2O \rightarrow 10SO_4^{2-} + 2MnO_4^- + 16H^+.$

(j) $2Fe^{3+} + 3C_2H_3OS^- + 3H_2O + 6NH_3 \rightarrow 2FeS + 3C_2H_3O_2^- + S^0 + 6NH_4^+.$

(k) $As_2S_3 + 14H_2O_2 + 12NH_3 \rightarrow 2AsO_4^{3-} + 3SO_4^{2-} + 12NH_4^+ + 8H_2O.$

(l) $Au^0 + 4Cl^- + 3NO_3^- + 6H^+ \rightarrow AuCl_4^- + 3NO_2 + 3H_2O.$

(m) $5Fe(SCN)^{2+} + 14MnO_4^- + 42H^+ \rightarrow 5Fe^{3+} + 14Mn^{2+} + 5SO_2 + 5CO_2 + 5NO_3^- + 21H_2O.$

(n) $2Mn^{2+} + 5BiO_3^- + 14H^+ \rightarrow 2MnO_4^- + 5Bi^{3+} + 7H_2O.$

(o) $2Hg_2^{2+} + NO_3^- + 4NH_3 + H_2O \rightarrow Hg_2O(NH_2)NO_3 + 2Hg^0 + 3NH_4^+.$

(p) $8Al^0 + 3NO_3^- + 5OH^- + 18H_2O \rightarrow 8Al(OH)_4^- + 3NH_3.$

(q) $2Fe(C_6H_5O_7)_2^{3-} + 4H_2S + 8NH_3 \rightarrow 2FeS + S_2^{2-} + 4C_6H_5O_7^{3-} + 8NH_4^+.$

(r) $Sb_2O_3 + 4Ag^+ + 4NH_3 + 2H_2O \rightarrow Sb_2O_5 + 4Ag^0 + 4NH_4^+.$

(s) $3PtCl_6^{2-} + C_3H_5(OH)_3 + 16OH^- \rightarrow 3Pt^0 + CO_3^{2-} + C_2O_4^{2-} + 18Cl^- + 12H_2O.$

(t) $3SCN^- + 13NO_3^- + 10H^+ \rightarrow 3SO_4^{2-} + 3CO_2 + 16NO + 5H_2O.$

(u) $2Pb^0 + 4C_2H_3O_2^- + O_2 + 4HC_2H_3O_2 \rightarrow 2Pb(C_2H_3O_2)_4^{2-} + 2H_2O.$

(v) $3HgS + 12Cl^- + 2NO_3^- + 8H^+ \rightarrow 3HgCl_4^{2-} + 2NO + 3S^0$
$+ 4H_2O.$

CHAPTER 8: Chemical Equivalence and Volumetric Stoichiometry

1. (a) $0.533F$ $CaCl_2 \cdot 6H_2O$, $0.533M$ Ca^{2+}, $1.07M$ Cl^-;
   (b) $3.34F$ $MgSO_4 \cdot 7H_2O$, $3.34M$ Mg^{2+}, $3.34M$ SO_4^{2-};
   (c) $0.30F$ $Ca(NO_3)_2$, $0.30M$ Ca^{2+}, $0.60M$ NO_3^-;
   (d) $0.060F$ Na_2SO_4, $0.12M$ Na^+, $0.060M$ SO_4^{2-};
   (e) $12.3F$ HCl, $12.3M$ H^+, $12.3M$ Cl^-;
   (f) $1.44F$ H_2SO_4, $2.88M$ H^+, $1.44M$ SO_4^{2-};
   (g) $3.34F$ HNO_3, $3.34M$ H^+, $3.34M$ NO_3^-;
   (h) $1.13F$ KOH, $1.13M$ K^+, $1.13M$ OH^-.
2. (a)   2.24 g KCl;
   (b)   2.01 g K_2SO_4;
   (c)   2.85 g $KAl(SO_4)_2 \cdot 12H_2O$;
   (d)   17.0 g $(NH_4)Al(SO_4)_2 \cdot 12H_2O$;
   (e)   16.7 g commercial NaOH;
   (f)   14.0 g $Ba(NO_3)_2 \cdot H_2O$;
   (g)   9.8 g $K_2Cr_2O_7$.
3. (a)   954 ml of 15.0 mg Cu^{2+}/ml;
   (b)   477 ml of $0.300F$ K_2SO_4;
   (c)   120 ml of $0.0500M$ NO_3^-;
   (d)   26.5 ml of $7.50F$ NH_3;
   (e)   41.7 ml of $HC_2H_3O_2$ (specific gravity 1.05, 35.0% weight solute);
   (f)   13.3 ml of 15.0 mg Cu^{2+}/ml;
   (g)   52.5 ml of $0.200F$ $BaCl_2$;
   (h)   260 ml of $0.0150N$ NO_3^-;
   (i)   50.0 ml of $0.800M$ Al^{3+}.
4. (a)   28.8 mmoles Cr^{3+}, 86.4 mequiv Cr^{3+}, 173 ml of $0.500N$ Cr^{3+};
   (b)   28.8 mmoles Cr^{3+}, 86.4 mequiv Cr^{3+}, 173 ml of $0.500N$ Cr^{3+};
   (c)   6.90 mmoles $Cr_2O_7^{2-}$, 41.4 mequiv $Cr_2O_7^{2-}$, 82.8 ml of $0.500N$ $Cr_2O_7^{2-}$;
   (d)   11.8 mmoles $Fe(CN)_6^{4-}$, 47.2 mequiv $Fe(CN)_6^{4-}$, 94.4 ml of $0.500N$ $Fe(CN)_6^{4-}$;
   (e)   11.8 mmoles $Fe(CN)_6^{4-}$, 11.8 mequiv $Fe(CN)_6^{4-}$, 23.6 ml of $0.500N$ $Fe(CN)_6^{4-}$;
   (f)   15.6 mmoles S^{2-}, 31.2 mequiv S^{2-}, 62.4 ml of $0.500N$ S^{2-};

(g)   15.6 mmoles S^{2-}, 31.2 mequiv S^{2-}, 62.4 ml of $0.500N$ S^{2-};

(h)   5.50 mmoles Mn^{2+}, 11.0 mequiv Mn^{2+}, 22.0 ml of $0.500N$ Mn^{2+};

(i)   5.50 mmoles Mn^{2+}, 5.50 mequiv Mn^{2+}, 11.0 ml of $0.500N$ Mn^{2+};

(j)   5.50 moles Mn^{2+}, 27.5 mequiv Mn^{2+}, 55.0 ml of $0.500N$ Mn^{2+};

(k)   0.619 mmoles $Co(NH_3)_6^{2+}$, 1.24 mequiv $Co(NH_3)_6^{2+}$, 2.48 ml of $0.500N$ $Co(NH_3)_6^{2+}$;

(l)   0.619 mmoles $Co(NH_3)_6^{2+}$, 0.619 mequiv $Co(NH_3)_6^{2+}$, 1.24 ml of $0.500N$ $Co(NH_3)_6^{2+}$.

5.   0.83 g Cu^{2+}, 0.013 mole Cu^{2+}, 0.026 g-equiv Cu^{2+}.

6.   0.53 ml of $2.5F$ NH_3.

7.   (a)   3.0 g Hg_2^{2+};
     (b)   1.0 g Hg_2^{2+};
     (c)   1.0 g Hg_2^{2+};
     (d)   1.7 g Hg_2^{2+};
     (e)   0.96 g Hg_2^{2+};
     (f)   4.0 g Hg_2^{2+}.

8.   20.3% $Al_2(SO_4)_3$.

9.   (a)   4.8 ml of $3.5M$ NO_3^-;
     (b)   6.2 ml of $3.5M$ NO_3^-;
     (c)   57 ml of $3.5M$ NO_3^-;
     (d)   1.5 ml of $3.5M$ NO_3^-;
     (e)   5.3 ml of $3.5M$ NO_3^-.

10. (a)   4.5 mequiv $KMnO_4$;
     (b)   0.14 g $KMnO_4$;
     (c)   0.90 mmole $KMnO_4$;
     (d)   3.0 ml of $0.30F$ $KMnO_4$.

11. 225 ml of $0.400F$ KI.

12. 0.29 g H_2O_2.

13. 3.9 g NO_3^-.

14. 66 ml of $1.2N$ Na_2SO_3.

15. $0.017N$ $SnCl_2$.

16. $0.022N$ $SnCl_2$.

17. (a)   0.250 mmole $Cr_2O_7^{2-}$;
     (b)   0.0540 g $Cr_2O_7^{2-}$;
     (c)   0.0735 g $K_2Cr_2O_7$;
     (d)   1.50 mequiv $K_2Cr_2O_7$;
     (e)   1.25 ml of $0.200M$ $Cr_2O_7^{2-}$;
     (f)   1.25 ml of $0.200F$ $K_2Cr_2O_7$;
     (g)   7.50 ml of $0.200N$ $K_2Cr_2O_7$;
     (h)   2.50 ml of $K_2Cr_2O_7$ solution.

18. $KMnO_4$ oxidizes 42.0 mequiv more and 3.77 g more of $Fe(OH)_2$.

19. 54.4% H_2SO_4.
20. 1.16 g HCl/ml.
21. 6.01F NaOH.
22. (a)   0.0240 g-formula AgCl;
     (b) Ag^+ in excess by 0.197 mole, 21.3 g, 0.197 g-equiv;
     (c)   0.133 mole Ba^{2+}/liter, 2.46 moles NO_3^-/liter.
23. 123 ml of H_2SO_4 solution.
24. 12.2 ml of NH_3 solution.
25. (a)   0.788N $KMnO_4$;
     (b)   0.0158F $KMnO_4$;
     (c) 5.12×10^{-4} g-equiv I^-;
     (d)   0.512 mequiv MnO_4^-.
26. 90.0 ml of 0.060F $AgNO_3$.
27. 0.45 g H_2S.
28. 0.46F H_2SO_4.
29. (a)   33.3 ml of 0.150F NaOH;
     (b)   66.7 ml of 0.150F NaOH;
     (c)   100 ml of 0.150F NaOH.
30. (a)   0.136 mole K^+/liter, 0.120 mole H^+/liter, 0.192 mole NO_3^-/liter,
            0.0320 mole SO_4^{2-}/liter;
     (b)   0.0640N SO_4^{2-};
     (c)   0.576N NO_3^-;
     (d)   0.600 g NaOH.
31. (a) 3.00×10^{-3} g-equiv I_2;
     (b) 1.00×10^{-3} mole Cr^{3+};
     (c)   0.240M I^-, 0.107M SO_4^{2-}, 0.0133M Cr^{3+}.
32. 4.8% acetic acid.
33. (a)   0.585 g $H_2C_2O_4$/100 ml;
     (b)   0.130N $H_2C_2O_4$;
     (c) i 8.13 mequiv $H_2C_2O_4$ and 0.602 g $H_2C_2O_4$,
            ii 16.3 mequiv $H_2C_2O_4$ and 0.602 g $H_2C_2O_4$.
34. 0.230N HCl, 0.230F HCl.
35. 38.3 g commercial KOH.
36. (a)   0.151F $Pb(NO_3)_2$, 0.302N $Pb(NO_3)_2$;
     (b)   0.151M Pb^{2+} and 0.302N Pb^{2+}, 0.302M NO_3^- and 0.302N NO_3^-;
     (c)   15.1 mequiv CrO_4^{2-}, 7.55 mmoles CrO_4^{2-}.
37. (a)   10.5 g K_2SO_4;
     (b)   5.23 g K_2SO_4;
     (c)   5.23 g K_2SO_4;
     (d)   5.23 g K_2SO_4;
     (e)   3.49 g K_2SO_4;
     (f)   52.3 g K_2SO_4;
     (g)   5.23 g K_2SO_4.
38. 16.3 ml of 2.00F $CaCl_2$, 81.7 ml of 1.60F HCl.

39.   341 ml of the NH_3 solution.
40.   89.2 g H_2O.
41.   333 ml.
42.   (a)   1.39 ml of $18.0F$ H_2SO_4;
      (b)   2.78 ml of $18.0F$ H_2SO_4;
      (c)   17.4 ml of $18.0F$ H_2SO_4;
      (d)   69.4 ml of $18.0F$ H_2SO_4.
43. 1.10% H_2O_2.
44.   (a)   6.4 g Cl_2;
      (b)   0.34 g Cl_2;
      (c)   0.45 g Cl_2;
      (d)   53 g Cl_2;
      (e)   53 g Cl_2;
      (f)   0.011 g Cl_2.
45.   (a) $0.0900N$ Mn^{2+}, $0.0450M$ Mn^{2+};
      (b) $0.146N$ Mn^{2+}, $0.0730M$ Mn^{2+};
      (c) $0.0920N$ Mn^{2+}, $0.0460M$ Mn^{2+};
      (d) $0.168N$ Mn^{2+}, $0.0336M$ Mn^{2+};
      (e) $0.750N$ Mn^{2+}, $0.150M$ Mn^{2+};
      (f) $0.108N$ Mn^{2+}, $0.0270M$ Mn^{2+}.
46.   (a)   24 ml of $0.25F$ $K_2Cr_2O_7$;
      (b)   40 ml of $0.25F$ $K_2Cr_2O_7$;
      (c)   81 ml of $0.25F$ $K_2Cr_2O_7$.
47.   (a) 80.5% purity;
      (b)   66.7 ml of solution.
48.   (a)   18.2 g MnO_4^-;
      (b)   4.28 g MnO_4^-;
      (c)   18.2 g MnO_4^-;
      (d)   10.9 g MnO_4^-;
      (e)   0.243 g MnO_4^-.
49.   (a)   2 mequiv H_3PO_4/mmole;
      (b)   0.0718 g HBO_3^{2-}.
50.   (a) $0.50N$ K_2CrO_4;
      (b) $0.75N$ K_2CrO_4;
      (c) $0.50N$ K_2CrO_4.
51.   (a)   indicator "T";
      (b)   indicator "N";
      (c)   indicator "Q";
      (d)   indicator "Q".
52.   77.9 g acid/g-equiv.
53.   47 ml of $0.040F$ $HC_2H_3O_2$.

APPENDIX B: METRIC STANDARDS OF MEASUREMENT

The following tabulations of familiar units of the metric (decimal) or cgs (centimeter-gram-second) system contain useful relationships of the weights and volumes encountered in text development — and, for the sake of completeness, some of length are also supplied in the thought that they prove of use.

1. Units of volume
   (a) *Cubic centimeter* (cc or cm³): this unit represents the volume capacity of a cube one centimeter on edge; hence, it fulfils a derivation based upon the fundamental metric unit of length, the *meter*.
   (b) *Liter*: the unit of capacity representing the volume occupied by one kilogram (1000 g) of H_2O at 3.98°C and 760 mm pressure.
2. Unit of mass
   Gram (g): This unit was originally standardized to represent the exact mass of one cubic centimeter of water at 3.98°C and 760 mm pressure. Subsequent refined measurements have revealed this unit to be actually 0.02 8 g heavier than originally warranted.
3. Unit of length
   Meter (m): This unit is standardized to and obtained by multiplying the wave length of the orange band of light in the spectrum of the element *krypton* of isotopic mass 86 by the factor 1,650,763.73.
   Prefixes are used to denote fractions or multiples of the standard units; thus among others:

   $deci$ = one-tenth; and $deka$ = ten.
   $centi$ = one-hundredth; and $cento$ = hundred
   $milli$ = one-thousandth; and $kilo$ = thousand
   $micro$ = one-millionth; and $mega$ = million

Inter-relationships of units and their parts
volume: 1 liter = 10 deciliters = 100 centiliters = 1000 milliliters = 1,000,000 (= 10^6) microliters
1 milliliter (ml) = 0.001 liter = 0.000001 (= 10^{-6}) kiloliter
1000 milliliters (ml) = 1 liter = 1000.28 cubic centimeters

mass: 1 gram (g) = 10 decigrams = 100 centigrams = 1000 milligrams = 1,000,000 (= 10^6) micrograms
1 milligram (mg) = 1000 micrograms = 0.001 gram = 0.000001 (= 10^{-6}) kilogram
also, 1000 (= 10^3) grams = 1 kilogram (kg)
and 0.000001 (= 10^{-6}) gram = 1 gamma

length: 1 meter (m) = 10 decimeters = 100 centimeters = 1000 milli-meters = 1,000,000 (= 10^6) micrometers or microns
1 millimeter (mm) = 1000 micrometers or microns = 0.001 meter = 0.000001 (= 10^{-6}) kilometers
1 angstrom (A) = 1×10^{-10} meter = 1×10^{-8} centimeter = 1×10^{-7} millimeter.

APPENDIX C: INTERNATIONAL ATOMIC WEIGHTS

Atomic weights are based on the isotope of carbon of mass number 12.0000...
The values in parentheses denote the mass number of the isotope of longest known half-life.

Name	Symbol	Atomic Number	Atomic Weight
Actinium	Ac	89	(227)
Aluminum	Al	13	26.9815
Americium	Am	95	(243)
Antimony	Sb	51	121.75
Argon	Ar	18	39.948
Arsenic	As	33	74.9216
Astatine	At	85	(210)
Barium	Ba	56	137.34
Berkelium	Bk	97	(249)
Beryllium	Be	4	9.0122
Bismuth	Bi	83	208.980
Boron	B	5	10.811
Bromine	Br	35	79.909
Cadmium	Cd	48	112.40
Calcium	Ca	20	40.08
Californium	Cf	98	(251)
Carbon	C	6	12.01115
Cerium	Ce	58	140.12
Cesium	Cs	55	132.905
Chlorine	Cl	17	35.453
Chromium	Cr	24	51.996
Cobalt	Co	27	58.9332
Copper	Cu	29	63.54
Curium	Cm	96	(245)
Dysprosium	Dy	66	162.50
Einsteinium	Es	99	(254)
Erbium	Er	68	167.26
Europium	Eu	63	151.96

Name	Symbol	Atomic Number	Atomic Weight
Fermium	Fm	100	(253)
Fluorine	F	9	18.9984
Francium	Fr	87	(223)
Gadolinium	Gd	64	157.25
Gallium	Ga	31	69.72
Germanium	Ge	32	72.59
Gold	Au	79	196.967
Hafnium	Hf	72	178.49
Helium	He	2	4.0026
Holmium	Ho	67	164.930
Hydrogen	H	1	1.00797
Indium	In	49	114.82
Iodine	I	53	126.9044
Iridium	Ir	77	192.2
Iron	Fe	26	55.847
Krypton	Kr	36	83.80
Lanthanum	La	57	138.91
Lawrencium	Lw	103	(257)
Lead	Pb	82	207.19
Lithium	Li	3	6.939
Lutetium	Lu	71	174.97
Magnesium	Mg	12	24.312
Manganese	Mn	25	54.9381
Mendelevium	Md	101	(256)
Mercury	Hg	80	200.59
Molybdenum	Mo	42	95.94
Neodymium	Nd	60	144.24
Neon	Ne	10	20.183
Neptunium	Np	93	(237)
Nickel	Ni	28	58.71
Niobium	Nb	41	92.906
Nitrogen	N	7	14.0067
Nobelium	No	102	(254)
Osmium	Os	76	190.2
Oxygen	O	8	15.9994
Palladium	Pd	46	106.4
Phosphorus	P	15	30.9738
Platinum	Pt	78	195.09
Plutonium	Pu	94	(242)
Polonium	Po	84	(210)
Potassium	K	19	39.102
Praseodymium	Pr	59	140.907
Promethium	Pm	61	(147)
Protoactinium	Pa	91	(231)
Radium	Ra	88	(226)
Radon	Rn	86	(222)
Rhenium	Re	75	186.2
Rhodium	Rh	45	102.905

Name	Symbol	Atomic Number	Atomic Weight
Rubidium	Rb	37	85.47
Ruthenium	Ru	44	101.07
Samarium	Sm	62	150.35
Scandium	Sc	21	44.956
Selenium	Se	34	78.96
Silicon	Si	14	28.086
Silver	Ag	47	107.870
Sodium	Na	11	22.9898
Strontium	Sr	38	87.62
Sulfur	S	16	32.064
Tantalum	Ta	73	180.948
Technetium	Tc	43	(99)
Tellurium	Te	52	127.60
Terbium	Tb	65	158.924
Thallium	Tl	81	204.37
Thorium	Th	90	232.038
Thulium	Tm	69	168.934
Tin	Sn	50	118.69
Titanium	Ti	22	47.90
Tungsten	W	74	183.85
Uranium	U	92	238.03
Vanadium	V	23	50.942
Xenon	Xe	54	131.30
Ytterbium	Yb	70	173.04
Yttrium	Y	39	88.905
Zinc	Zn	30	65.37
Zirconium	Zr	40	91.22

APPENDIX D: PERIODIC TABLE OF THE ELEMENTS

Group	I	II				Transition Elements								Post-Transition Elements		III	IV	V	VI	VII	0
			III	IV	V	VI	VII		VIII			I	II								
Period 1	H 1																				He 2
2	Li 3	Be 4													B 5	C 6	N 7	O 8	F 9	Ne 10
3	Na 11	Mg 12													Al 13	Si 14	P 15	S 16	Cl 17	Ar 18
4	K 19	Ca 20	Sc 21	Ti 22	V 23	Cr 24	Mn 25	Fe 26	Co 27	Ni 28	Cu 29	Zn 30			Ga 31	Ge 32	As 33	Se 34	Br 35	Kr 36
5	Rb 37	Sr 38	Y 39	Zr 40	Nb 41	Mo 42	Tc 43	Ru 44	Rh 45	Pd 46	Ag 47	Cd 48			In 49	Sn 50	Sb 51	Te 52	I 53	Xe 54
6	Cs 55	Ba 56	La* 57	Hf 72	Ta 73	W 74	Re 75	Os 76	Ir 77	Pt 78	Au 79	Hg 80			Tl 81	Pb 82	Bi 83	Po 84	At 85	Rn 86
7	Fr 87	Ra 88	Ac▾ 89																	

Lanthanides *58–71	Ce 58	Pr 59	Nd 60	Pm 61	Sm 62	Eu 63	Gd 64	Tb 65	Dy 66	Ho 67	Er 68	Tm 69	Yb 70	Lu 71
Actinides ▾90–103	Th 90	Pa 91	U 92	Np 93	Pu 94	Am 95	Cm 96	Bk 97	Cf 98	Es 99	Fm 100	Md 101	No 102	Lw 103

INDEX